The Mathematics of Options

Michael C. Thomsett

The Mathematics of Options

Quantifying Derivative Price,
Payoff, Probability, and Risk

Michael C. Thomsett
Spring Hill
TN, USA

ISBN 978-3-319-85960-6   ISBN 978-3-319-56635-1 (eBook)
DOI 10.1007/978-3-319-56635-1

© The Editor(s) (if applicable) and the Author(s) 2017
Softcover reprint of the hardcover 1st edition 2017
This work is subject to copyright. All rights are solely and exclusively licensed by the Publisher, whether the whole or part of the material is concerned, specifically the rights of translation, reprinting, reuse of illustrations, recitation, broadcasting, reproduction on microfilms or in any other physical way, and transmission or information storage and retrieval, electronic adaptation, computer software, or by similar or dissimilar methodology now known or hereafter developed.
The use of general descriptive names, registered names, trademarks, service marks, etc. in this publication does not imply, even in the absence of a specific statement, that such names are exempt from the relevant protective laws and regulations and therefore free for general use.
The publisher, the authors and the editors are safe to assume that the advice and information in this book are believed to be true and accurate at the date of publication. Neither the publisher nor the authors or the editors give a warranty, express or implied, with respect to the material contained herein or for any errors or omissions that may have been made. The publisher remains neutral with regard to jurisdictional claims in published maps and institutional affiliations.

Cover design: Fatima Jamadar

Printed on acid-free paper

This Palgrave Macmillan imprint is published by Springer Nature
The registered company is Springer International Publishing AG
The registered company address is: Gewerbestrasse 11, 6330 Cham, Switzerland

Foreword

Writing a book—any book—looks easy after it is completed. However, the process itself is far from simple. An author must remember the proper tone for a target audience and decide what to include and exclude in every chapter. Finally, the author has to make sure that the essential message is communicated both clearly and accurately.

I have struggled in my own career with the challenges of communication. This is an unending process in my current roles as lecturer of energy trading and risk management for Fitch Learning's *Certificate in Quantitative Finance Programme* at Wall Street, energy trading lecturer at the Freeman School of Business at Tulane University in New Orleans, and author of three books, and in my development of www.MathQED.com—a mathematics homework platform for K-12 and college students, teachers, and parents.

Just as I am gratified when I see a student's face lights up at a realization of an underlying concept, teaching energy trading and applied mathematical topics at every level involves bridging that gap between theory and real-world applications. *The Mathematics of Options* accomplishes the goal of leading the reader through a maze of concepts and terminology to arrive at a rather simple result, the appreciation of concepts needed to become a successful options trader.

Too often, the mathematical aspects of options are made overly complex and unnecessarily obscure. Emphasis on theoretical ideas, such as options pricing models or estimated future volatility is of no use to a trader who needs a few basic tools: An understanding of risk and probability, a means to visualize payoffs and volatility, and identification of a range of possible trading outcomes. In Thomsett's book, risk and probability are expressed in practical terms. Historical volatility is explained and demonstrated as an effective means for timing trades. In addition, the outcomes (profit, breakeven, or loss) are summarized mathematically and illustrated so that every trader can simplify the process of selecting trades, rather than complicating it.

One truth I learned when I wrote my second book, *Energy Trading and Risk Management: A Practical Approach to Hedging, Trading and Portfolio Diversification* (John Wiley & Sons, 2014), is that an outside observer might assume that writing about options, hedging, and diversification comes easily to an expert. It does not! Even with a Harvard Ph.D. in Applied Mathematics, a London Business School Executive MBA, and experience as a faculty member at MIT, I discovered a challenge in articulating what readers needed to know about financial and energy derivatives. It does not matter how much knowledge the author possesses, or how well the author can communicate knowledge through the written word.

With this in mind, I applaud any author who is also an effective teacher, and who can explain the essential message in a book. *The Mathematics of Options* tackles a very big topic, indeed, and identifies not only what an options trader or professional needs to know, but also what can be rejected as nonessential. Thomsett's book demonstrates what works on a practical level, and what does not work. That makes this unusual, not only as a book of mathematics but also as a book of options trading.

Having worked at several universities, energy and financial institutions in the USA, London, and Asia, I have seen a repetitive challenge in each of my work environments. Complexity does not produce results; simplicity does. We need to strip education back to basics. Math, especially, too often is not taught effectively and, as a consequence, many intelligent and capable students are left behind when energetic and effective educational methods would have made all the difference. Taking on the challenge of writing a mathematics book—especially one involving the options world—is a daunting task, but not impossible, as Thomsett's book demonstrates.

Challenges, in fact, define everyone's degree of success in their profession. As a derivatives and applied math educator from very humble beginnings, I have discovered a great truth about how the world works: Math, such as reading, opens many doors. It is the gatekeeper subject to so many interesting and even lucrative professions, such as engineering, science, technology, computer science, trading, medicine, etc. Hence, sadly, we limit ourselves *only* by what we fail to pursue.

For anyone interested in improving their knowledge of math and options trading, the only limitation to progress is internal. This book, in its ease of communication of a complex topic, results from Thomsett having identified the target audience, sets the appropriate tone, and precisely determined the information to include or exclude.

New Orleans, USA
March 2017

Iris Mack, Ph.D.

Preface

The Misunderstood Options

Options trading may have vastly different appearances, depending on the observer. The speculator treats options as efficient forms of leverage, and the conservative portfolio manager or equity investor seeks hedging through options to manage, reduce, or eliminate risk.

This book presents a range of information about specific groupings of strategies, all quantified mathematically in terms of profit, breakeven, or loss. The concept here is that even the most sophisticated and experienced options trader is likely to benefit from awareness of these all-important benchmarks for every trade. In setting price or profit goals for exit, these calculations are especially useful. Every trader is able to set goals for taking profits or accepting losses. That is the easy party. Actually taking action when those levels are reached is far more difficult.

Beyond the highly detailed chapters illustrating the appearance of profit and loss for many types of trades and applying these formulas, the book contains many additional chapters of interest. Among the observations in the book is a questioning of implied volatility as a useful tool. As a measurement of probability for favorable outcomes, historical volatility may offer an equal or better yardstick for identifying risk and opportunity. This is a topic of passionate debate, and both sides—those who believe in implied volatility and those who reject it—have their rationale.

Another observation in the book questions the methods used in the market for calculating probability. In Chap. 1, the comparison between two separate methods is traced back to a chronic gambler in 175th-century France, the Chevalier de Mere. He realized that his carefully constructed probability of favorable outcomes was not being matched in actual play. He consulted with his friend, the famous Blaise Pascal who, in turn, consulted with Pierre de Fermat, a gifted mathematician. These two discovered that probability of getting a favorable outcome was *not* accurate. The more reliable system was to first calculate the probability of not getting the favorable outcome, and then subtracting the result from one.

This paradox—the difference between additive and multiplicative probability—is profound and has importance in modern options trading. So many examples of what is broadly lumped in together under the umbrella of "probability" involve the less reliable additive method of calculation. With this in mind, probability itself has to be studied by options traders not as a system of guarantees, but more as a system for better understanding the potential within a range of expected outcomes.

Just as de Mere was a chronic gambler in his day, many options traders are equally attracted to the law of averages. But many are further puzzled because outcomes do not always conform to expectations. Albert Einstein cautioned that in his opinion, God does not play dice. This is not a rejection of *risk* by any means, but an observation that probability involves only estimates of possible outcomes. This often is misunderstood by traders who find comfort in applying an online calculator and discovering that the odds of a particular options ending up in the money are 80%. But how is that probability calculated? The online source does not reveal its methods nor its assumptions, so traders are expected to grant significant trust in a source that does not disclose any details.

These examples of topics and their treatment within this book are only part of a broader set of assumptions concerning what may be termed *practical* mathematical application. This means that formulations involving a lot of theory and the exponential uncertainties of multiple variables are not used to set a basis for the premise underlying the book: Traders want to know the levels of profit, breakeven, and loss, and they want to be able to quantify risk. It is that simple.

On a theoretical level, much has been written about the wonders of pricing models, especially the best-known among these, the Black-Scholes pricing model. This book attempts to base a system on historical volatility, technical analysis, and fundamental analysis as the means for selecting underlying securities and their options. But why reject Black-Scholes? Actually, there are numerous reasons. Fischer Black himself wrote several years after publication of the original formula that the assumptions used by himself and Myron Scholes were, in fact, deeply flawed. He identified *nine* specific flaws and incorrect assumptions, each of which distorts the pricing model. A single flawed assumption is concerning enough, but may be accepted as part of an analytical process. However, nine flaws are exponentially more serious and bring into question the entire process of developing a pricing model. This led to an equally important question: Why do options traders need a pricing model at all?

There are no pricing models for other forms of investing or trading. A trader, speculator, or investor focusing on stocks, for example, relies on the market and its forces of supply and demand to set prices as a floating and ever-changing aspect of that auction market, and it works quite well. The market, with its informational efficiency at play, sets up a universe in which astute traders recognize when securities are overpriced or underpriced, all without a pricing model or formulation of what the price per share *should* be.

Since options are derivatives of their underlying securities, the pricing of an options contract is directly derived from movement in the underlying price, which varies and changes based on both price movement and historical volatility. The conclusion: A pricing model is a comfortable concept in theory, as Fischer Black himself observed as a certainty that could provide comfort to traders. However, Black also said, the assumptions that go into this desire for perfection rarely can be applied so that the pricing model works.

With these controversial realizations and the natural conflict between a comforting theory and a stark reality, it becomes clear that no one's answer addresses all of the questions or satisfies all of the beliefs within the options universe. By presenting alternatives, individuals on both sides may enter into a debate and may even learn from one another. Unfortunately, disagreement too often leads to shutting out of the opportunity to expand knowledge. It may only be hoped that the information in this book, especially surrounding these controversial topics, will lead to an advanced appreciation of a complex topic, whether through confirmation of what is believed today, or by discovery of alternative possibilities. That process—learning how to view a problem from a different point of view—is how everyone learns.

Spring Hill, USA   Michael C. Thomsett

Contents

1	Trading Goals and Objectives	1
2	The Role of Fundamental and Technical Analysis	31
3	Pricing of the Option	55
4	The Dividend Effect	79
5	Return Calculations	99
6	Strategic Payoff: The Single-Option Trade	125
7	Strategic Payoff: Spreads	161
8	Strategic Payoff: Straddles	197
9	Probability and Risk	231
10	Option Pricing Models	255
11	Alternatives to Pricing Models	269

Appendix—Formulas 285

Bibliography 319

Index 325

List of Figures

Fig. 1.1	Tiffany—chart courtesy of StockCharts.com	10
Fig. 1.2	The bell curve—prepared by the author	12
Fig. 1.3	Standard deviation—prepared by the author	14
Fig. 1.4	Alamo group—chart courtesy of StockCharts.com	20
Fig. 1.5	Matson—chart courtesy of StockCharts.com	21
Fig. 1.6	IBM—chart courtesy of StockCharts.com	22
Fig. 1.7	J.M. Smucker—chart courtesy of StockCharts.com	23
Fig. 1.8	Total options contracts by volume—prepared by the author	27
Fig. 2.1	JC Penney—chart courtesy of StockCharts.com	34
Fig. 2.2	Price and revenue history, Wal-Mart—chart courtesy of StockCharts.com	43
Fig. 2.3	Price and revenue history, J.C. Penney—chart courtesy of StockCharts.com	43
Fig. 2.4	Wal-Mart stock chart—chart courtesy of StockCharts.com	48
Fig. 2.5	Wal-Mart stock chart, expanded—chart courtesy of StockCharts.com	50
Fig. 2.6	J.C. Penney stock chart—chart courtesy of StockCharts.com	50
Fig. 3.1	Moneyness of an option—chart courtesy of StockCharts.com	65
Fig. 3.2	Time decay—prepared by the author	66
Fig. 3.3	Gamma and the effect of volatility—prepared by the author	73
Fig. 3.4	Priceline stock chart—chart courtesy of StockCharts.com	75
Fig. 3.5	Intuitive Surgical price chart—chart courtesy of StockCharts.com	76
Fig. 4.1	10-Year price chart comparisons—chart courtesy of StockCharts.com	85
Fig. 4.2	Two-company price chart comparisons—chart courtesy of StockCharts.com	88

Fig. 5.1	J.P. Morgan Chase (JPM), 3 years—chart courtesy of StockCharts.com	121
Fig. 5.2	ConocoPhillips (COP), 3 years—chart courtesy of StockCharts.com	122
Fig. 6.1	Payoff, long call—prepared by the author	133
Fig. 6.2	Payoff, long put—prepared by the author	135
Fig. 6.3	Payoff, uncovered ITM call—prepared by the author	139
Fig. 6.4	Payoff, uncovered OTM call—prepared by the author	142
Fig. 6.5	Payoff, uncovered put—prepared by the author	144
Fig. 6.6	Amazon.com (AMZN) price chart—chart courtesy of StockCharts.com	148
Fig. 6.7	Payoff, covered call—prepared by the author	152
Fig. 6.8	Payoff, ratio call write—prepared by the author	153
Fig. 6.9	Payoff, variable ratio call write—prepared by the author	155
Fig. 6.10	Payoff, covered put—prepared by the author	158
Fig. 7.1	Payoff, bull credit spread—prepared by the author	165
Fig. 7.2	Payoff, bull call debit spread—prepared by the author	167
Fig. 7.3	Payoff, bear put debit spread—prepared by the author	169
Fig. 7.4	Payoff, bear call credit spread—prepared by the author	171
Fig. 7.5	Payoff, condor—prepared by the author	174
Fig. 7.6	Payoff, iron condor—prepared by the author	176
Fig. 7.7	Payoff, long butterfly—prepared by the author	178
Fig. 7.8	Payoff, short butterfly—prepared by the author	180
Fig. 7.9	Payoff, iron butterfly—prepared by the author	182
Fig. 7.10	Payoff, synthetic long stock—prepared by the author	185
Fig. 7.11	Payoff, synthetic short stock—prepared by the author	187
Fig. 7.12	Payoff, bull calendar spread—prepared by the author	190
Fig. 8.1	Payoff, long straddle—prepared by the author	203
Fig. 8.2	Payoff, short straddle—prepared by the author	205
Fig. 8.3	Payoff, covered straddle—prepared by the author	209
Fig. 8.4	Payoff, long strangle—prepared by the author	212
Fig. 8.5	Payoff, short strangle—prepared by the author	214
Fig. 8.6	GOOG price chart—chart courtesy of StockCharts.com	215
Fig. 8.7	Payoff, long gut strangle—prepared by the author	218
Fig. 8.8	Payoff, short gut strangle—prepared by the author	220
Fig. 8.9	Payoff, long strip—prepared by the author	224
Fig. 8.10	Payoff, long strap—prepared by the author	226
Fig. 8.11	Alphabet (GOOG) with BB expansion—chart courtesy of StockCharts.com	227
Fig. 9.1	Positive and negative skew—prepared by the author	233
Fig. 9.2	Historical volatility, consolidating range—chart courtesy of StockCharts.com	235
Fig. 9.3	Historical volatility, trending range	236

List of Tables

Table 1.1	Excel formula, calculation of standard deviation—prepared by the author	16
Table 2.1	Excel formula, calculation of historical volatility—prepared by the author	36
Table 2.2	Fundamental outcomes, 10 years—prepared by the author	40
Table 2.3	Change comparisons, WMT and JCP—prepared by the author	42
Table 2.4	Rating system for proximity trade timing—prepared by the author	47
Table 2.5	Rating system, Wal-Mart—prepared by the author	49
Table 2.6	Rating system, J.C. Penney—prepared by the author	51
Table 3.1	Approximation of delta for long options—prepared by the author	69
Table 3.2	Approximation of delta for short options—prepared by the author	71
Table 4.1	10-Year debt capitalization ratio comparison–prepared by the author	83
Table 4.2	Initial annualized total return—prepared by the author	89
Table 4.3	Expanded annualized total return—prepared by the author	90
Table 4.4	Dividend yield—prepared by the author	94
Table 5.1	Breakeven rate of return—prepared by the author	103
Table 5.2	Covered call total return comparisons—prepared by the author	105
Table 6.1	Long call premium comparisons, Boeing (BA)—prepared by the author	132
Table 6.2	Long put premium comparisons, Boeing (BA)—prepared by the author	135

Table 6.3	Short call ITM comparisons, Apple (AAPL)—prepared by the author	139
Table 6.4	Short call OTM comparisons, Tesla (TSLA)—prepared by the author	141
Table 6.5	Short put comparisons, Tesla (TSLA)—prepared by the author	144
Table 6.6	Effect of earnings surprise, Amazon (AMZN)—prepared by the author	147
Table 6.7	Covered call comparisons, Netflix (NFLX)—prepared by the author	150
Table 6.8	Covered put comparisons, Apple (APPL)—prepared by the author	157
Table 7.1	Calls and puts, Southwest airlines (LUV)—prepared by the author	164
Table 7.2	Calls and puts, Bristol-Myers Squibb (BMY)—prepared by the author	168
Table 7.3	Calls and puts, Macy's (M)—prepared by the author	173
Table 7.4	Calls and puts, Occidental Petroleum (OXY)—prepared by the author	177
Table 7.5	Calls and puts, Kellogg (K)—prepared by the author	184
Table 7.6	Calls and puts, MGM Resorts (MGM)—prepared by the author	190
Table 7.7	Calls and puts, Chipotle (CMG)—prepared by the author	193
Table 8.1	Calls and puts, Anheuser-Busch (BUD)	202
Table 8.2	Calls and puts, Alphabet (GOOG)	211
Table 8.3	Calls and puts, General Mills (GIS)	217
Table 8.4	Calls and puts, Intuitive Surgical (ISRG)	223

Introduction—The Variability of Derivatives Trading

Abstract

Professional traders need practical solutions combining essential mathematical principles used to quantify options value. This should be practical and actionable, so that options traders are able to make clear and informed decisions. Every options trader deals with an array of calculations. This applies to pricing, of contracts, payoff expectations of specific strategies (including maximum profit and loss), probability of outcomes, and identification of risks. Traders want a convenient and practical reference guide to hedging portfolio risk; for evaluating pricing, payoff, probability and risk issues; and to convert complexity into sensible decisions. Among the challenges to articulating mathematical quantification of options is the problem of inaccuracies and estimates as part of so many formulas. Traders need and deserve a straightforward and accurate way to measure the essential ingredients of trading.

Every options trader deals with an array of calculations. This applies to pricing, of contracts, payoff expectations of specific strategies (including maximum profit and loss), probability of outcomes, and identification of risks.

The beginner is concerned with identifying risk and opportunity and focusing on a short list of strategies. However, the experienced options trader is more focused on a broad range of issues. Among these is the need for hedging to reduce portfolio risk as one aspect of options trading. The experienced options trader requires reference for evaluating these pricing, payoff, probability, and risk issues. These convenient and common sense calculations are not easily found in free online sources, which tend to fall into one of two categories: First is the basic type of calculation aimed at the novice or general trader. Second is the esoteric level of advanced probability analysis and the Black-Scholes pricing model, aimed at the theoretical rather than at practical applications.

For the practical use of options math, notably for experienced individual traders or investors and for portfolio managers, the literature of the industry has lacked a practical reference. This is why this book has been written. Among the many challenges to articulating mathematical quantification of options trading is the problem of inaccuracies and estimates as part of so many formulas. One inaccuracy in a calculation is troubling, but can be adjusted in order to approximate a reliable outcome. However, multiple inaccuracies create exponential levels of unreliability.

A second problem with many mathematical applications is the lack of clarity in the methods used to calculate outcomes, such as probability, implied volatility, and pricing estimates. If traders are to rely on spreadsheets or calculators that include variables, how much reliance should be placed in the outcome? If the brokerage firm of options service is not able to willing to disclose its assumptions, the reliability of a simplified calculator cannot be known. This presents a deep and chronic problem for anyone interested in determining a reliable set of prices, risks, or trading opportunities.

The book begins with a review of some of the basics, as a means for setting the tone of the book and for defining a starting point for the mathematical principles addressed in the book. Chapter 1 is a discussion of trading goals and objectives, meant to reiterate the differences between investing and speculating and how options fit within that broad spectrum of risk tolerance. It also matches the specific risk tolerance attributes with risk identification specific to a range of options strategies. In order to identify the means for calculating expected outcomes based on well-defined goals and objectives, the chapter proposes development and use of a "probability matrix" designed to create a visual summary of the mathematical risk parameters associated with stock price behavior and, by association, of ever evolving options values and risks.

Chapter 2 further examines the rudimentary aspects of any trading program by exploring the role of fundamental and technical analysis for options trading. The concept of using both forms is not often addressed, neither is it common to suggest that both are of great value. However, as a starting point in determining which companies to include in an options-based program, the fundamental strength or weakness is an essential starting point to develop a sound program, not only for the obvious equity selection process but also to determine risk profiles for options strategies.

In Chap. 3, the pricing of options is the topic. This might seem an obvious and basic attribute of options trading, but in fact the methods by which premium levels are judged may involve a range of considerations including put/call parity and upper/lower price bounds.

Chapter 4 describes how dividends affect options returns and how a various number of dividend-related calculations have to be considered in the mathematical analysis of options trading.

It might appear that calculating returns on options trading is exceptionally basic. In practice, however, this may be one of the more complex aspects to options and often involves inconsistencies and confusion, even among experienced traders. Chapter 5 describes the problems and offers solutions in developing a consistent and reliable program for accurate comparisons and analysis of likely outcomes.

Chapter 6 describes the various forms of single-options trades, including meaning the most basic long calls and puts as well as a variety of short option trades. The chapter examines the critical importance of proximity as a timing issue for covered calls. Beyond the single-options trade is the popular spread in its many forms. Chapter 7 demonstrates the strategic payoff calculations of spread strategies, including proximity and its risk factors that affect pricing for the underlying security.

In Chap. 8, the same attributes are examined for straddles as alternative strategies—both high-risk and conservative—and further demonstrates how the math in quantifying risk is essential to understanding these devices. Chapter 9 describes the probability and risk of options trading, expanding to development of a probability matrix, use of VaR, expected and unexpected loss, and an overview of the mathematics of risk tolerance.

In Chap. 10, pricing models are examined and dissected. This chapter includes an analysis of the popular Black-Scholes pricing model and explores its benefits as well as its flaws. Moving beyond a discussion of pricing models, Chap. 11 demonstrates the many alternatives available to the options professional. Those who recognize the problems associated with reliance on pricing models, the Greeks, and calculations of implied volatility, will find practical value in reliance on technical timing, proximity analysis, and signals found in the forms of price, volume, momentum, and moving average analysis.

The intention and purpose of this book is to offer practical ideas for professional traders combining essential mathematical principles to quantify options value, with a practical and actionable approach to the broader struggle of identifying how to make informed decisions. Trade entry and exit timings are at the core of this effort, and options traders, such as everyone in the market, seek solutions to the timing challenges in an uncertain environment. The usual statistical analyses used elsewhere are rarely applicable. In most forms of statistics, a fixed field of known variables is involved. In the markets, however, the field is constantly in flux, changing not only daily but also by the minute. As a consequence, the random variables of the market necessitate an approach that moves beyond the comforting statistical certainty and expand to present a fresh and informed field of vision that acknowledges the dimensions of variation every trader faces.

1
Trading Goals and Objectives

Chapter Objectives:

– define investing objectives based on appreciation of risk
– evaluate the value of implied volatility to price options
– analyze whether volatility follows or leads price
– consider option strike proximity in timing trades
– develop a probability matrix to visualize volatility
– apply indicators like Bollinger Bands to improve timing of option trades.

Here is the challenge every trader and investor faces: Upon defining the goals and objectives for the use of options, how can you next develop a simple but effective mathematical model to (a) ease the burden of calculation, (b) add confidence to the timing to entry or exit, and (c) ensure that self-imposed risk tolerance levels are adhered to in the decision making process?

This is a rather large challenge. For many, relying on complex theories and pricing models, it may be overwhelming, and the complexity yields no actionable answers. Excessive reliance on estimates and multiple variables is a flawed methodology. But there is a solution. The math itself can be reduced to a visual signaling system enabling you to identify probability at a glance and to improve timing of trades, all without needing to convert a trading methodology to a speculative model. The speculator, in the analysis of these challenges, has always defined *risk* associated with options trading, but that is not necessarily the end of the story. In fact, applying a visual adaptation of probability analysis reduces risks while helping investors and traders to spot strong reversal and confirmation signals.

Speculators have long employed options as a favored tool for what is doubtlessly a welcome acceptance of risk. The idea is that by taking bigger risks, bigger profits follow. However, even the most devoted speculator also has to acknowledge that with big profit opportunities come equal (or greater) risk of big losses. This is why every trader or investor using options for more conservative applications (notably hedging equity portfolio risk exposure) needs to set specific goals and objectives, and to develop the self-discipline to adhere to these based on well-defined risk tolerance levels.

The "goal" defines policies regarding options trades. How much is placed at risk, and for how long? When will positions be closed to accept profits or losses, and is this based on dollar values or percentages of gain or loss? The "objective" is a clarification of the purpose behind using options. Are these short-term speculations, intentionally limited in dollar-value exposure? Or are the options trades a core form of leverage aimed hedging equity risks and out-performing the market?

Most traders and investors including options as part of a portfolio appreciate the distinction between goals and objectives. However, the definition itself easily is abandoned when it comes to entering trades and determining if or when to exit. Many self-defined "conservative" traders regard options as tools for hedging equity risks. However, by the nature of how they trade and the types of strategies employed, they tend to speculate more often than not. If this is acceptable to you as a trader or investor, there is no flaw as long as the risk levels associated with speculation are well articulated and adopted as part of the overall strategic use of options. However, if your goals and objectives define a preference for conservative strategies, then the practices should be reviewed carefully and thoroughly.

This requires, as part of developing an options-based program, that the degree of speculative behavior deemed acceptable is thoroughly defined as an initial phase of defining how to trade, and what trading strategies will be used to accomplish a fully articulated set of goals and objectives. In order to identify how much speculation is appropriate with options, given your risk tolerance (and within the restrictions of goals and objectives), the nature of strategies and their relative simplicity or complexity determines how effectively those goals and objectives may be met. This also is where so many investors and traders are likely to discover a stark reality, that their self-defined level of risk tolerance is violated, perhaps on a daily basis.

In operating as a speculator, a distinction should also be made between profit desirability and risk transfer. These two related but distinct attributes of speculation often are overlooked in favor of merely finding an options trade deemed to be "acceptable" on some standard, often poorly defined. A frequently used model for risk transfer is the Keynes-Hicks theory, which "may be called the *risk-transfer hypothesis*. On this view, speculators are relatively risk-tolerant individuals who are rewarded for accepting price risks from more risk-averse hedgers."[1]

Moving this definition forward, "… the *Keynes-Hicks theory* of speculation emphasizes not differences in beliefs, but differences in willingness to take risk or in initial positions as the foundation of a speculative market."[2]

This has direct implications for options trading, whether used speculatively or for hedging the equity portfolio. It is not merely the willingness to accept loss, but the perception of risk itself that defines the speculator. For many options traders, an informed evaluation of specific strategies may easily lead to a conclusion that risks are low. However, an unhealthy dose of confirmation bias may in fact lead to ill-advised trades. Thus, the self-defined conservative options trader might also suffer from a well-developed and informed understanding of the options market, in the sense that the extent of the true risks is poorly understood or easily overlooked. The desirability of a conservative portfolio management policy is easily replaced with an excessively speculative set of practices. This returns the discussion to one of setting and adhering to well-defined goals and objectives. If you define yourself as conservative, your goals and objectives reflect that definition, but does your trading pattern also conform? All too often, it will not. Thus, risk itself must be quantified as part of this self-defining process.

Basic Probability to Quantify Risk

Ironically, options trading involves the use of probability as a means for identifying timing of trade entry and exit, often with the use of online free calculators. However, the question has to be asked: Does the options trader understand the relationship between "probability" and risk avoidance? Putting this another way, what are the chances that analysis of a probability of success is not enough? It may be that moving beyond the initial understanding of probability outcomes calculated on a basis statistical model is less accurate than the analyst assumes.

Applying a mathematical formula to quantify risk often is elusive. This is why clarification of options risk is best accomplished by first defining the probability of profit or loss. This does not mean that esoteric pricing formulas are the answer. To the contrary, pricing models do not address the core question, how much risk is involved in a specific option trade? Augmenting the ineffectiveness of pricing models, the probability calculation itself may not reveal an accurate enough answer to the question every trader has: *What is my probability of success?*

A more detailed examination of risk points to the potential for misapplication of probability in determining whether or not to enter an options trade. Relying on the model for potential success versus potential failure makes the point about probability and its application to options trading.

An example based on finite possible outcomes clarifies the problem options traders face regarding probability. The roll of a single die appears to contain a 67% chance of any one result occurring in a series of four rolls. This statistical assumption is easily perceived by first recognizing the one-in-six chance in a single roll of any one number resulting. For four rolls, the probability formula for this is:

$$4 * 1/6 = 67\%$$

You know there is a one-in-six chance because there are six possible results—1, 2, 3, 4, 5, and 6. However, when two dice are employed, the calculation of probable outcomes is more complex. For example, an initial calculation using the same formula as above yields a similar result if the number of rolls is expanded. The factor used for two dice requires multiplying the one-die factor by itself:

$$1/6 * 1/6 = 1/36$$

You can prove the total of 36 by listing all possible combinations:

1-1	1-4	3-3	4-3	5-3	4-6
1-2	2-3	4-2	5-2	6-2	5-5
2-1	3-2	5-1	6-1	3-6	6-4
1-3	4-1	1-6	2-6	4-5	5-6
2-2	1-5	2-5	3-5	5-4	6-5
3-1	2-4	3-4	4-4	6-3	6-6

An application of the probability formula assuming 24 rolls is:

$$24 * 1/36 = 67\%$$

However, as evident as this formula appears, it is not accurate. To determine the true probability of any one outcome, it is necessary to calculate the probability of that outcome *not* occurring. Because there are many more chances of failure, the additive method above is not a reliable or accurate probability result. Thus, the odds of rolling any number with a single die are not accurate at 67%, but rather produce a result of 52%. This is apparent when you calculate the odds of any one number not being rolled in 4 attempts, and then subtracting that total from 1:

$$1-(5 \div 6)^4 = 52\%$$

This is dramatically different than the initial calculation, revealing that with the odds of a number coming up are 52%, and not 67%.

Applying this to two dice with 24 rolls, a different and lower outcome is arrived at when the calculation determines the odds, lower in fact than for the single die calculation. The odds of a specific two-dice outcome not occurring in a series of 24 rolls and then subtracting from 1:

$$1-(35 \div 36)^{24} = 49\%$$

The fact that in 24 rolls, the probability of any one outcome resides at 49% rather than at 67%, vastly changes the probability, thus the risk level. The initial calculation of a single value coming up 67% of the time out of 4 rolls, or a double value in 24 rolls, is inaccurate when the more advanced multiplicative formula is applied.

This insight was first provided in the seventeenth century, when a collaboration between Blaise Pascal and Pierre de Fermat developed the understanding of probability. This was the advent of modern-day probability theory.[3]

With options trading, the same thinking applies to more accurately reveal the success of a specific strategy. Options traders tend to think of their trades not so much for hedging value, but in how a speculative result can be devised to "beat the market" and develop a consistent method for generating profits. A realistic assessment of risks requires a process similar to that of calculating probability for dice rolls. If an options trader believes that profits are likely at the rate of 67% in 'x' trades, when the true probability is closer to a 50/50 result, the implications are clear. Such a trader may enter into a series of traders assumed to contain the same risk attributes, expecting to profit at a level far greater than actual probability. Although the example of dice is narrow and finite, the same observation applies for the greater level of variables involved in options trading.

The process of defining goals and objectives, more than anything else, is a matter of articulating levels of risk. Some are acceptable and some are not. The difficult part is defining which is which, and a sincere effort at this reveals that the widely known term "risk tolerance" is more complex than a mere list of strategies that are deemed as "good" or "bad" under an initial definition. In fact, risk tolerance cannot be truly defined until a range of revealing signals are brought to bear, in order to articulate the true risks in an investing or trading program.

The Flaws of Implied Volatility (IV)

In observing technical patterns and probability that options traders may consider replacing the popular analysis of implied volatility (IV) with a visual form of probability analysis. This means that the timing of options trades often is more effective when timed to coincide with signals in the underlying security. Of course, other factors have to come into play in the selection and timing of an options strategy, including moneyness of the option, time remaining until expiration, underlying price (historical) volatility, and proximity of current underlying price to resistance or support. However, the point worth making here is that the math of the underlying security may be a more valued form of timing intelligence, than timing based solely on the option price or implied volatility.

Why not rely on implied volatility (IV)? There are several reasons to move away from IV as the default method for timing of trades. The general belief is that high volatility points to timing for entering short positions or closing long ones; whereas low volatility is the timing mechanism for entering long and closing short positions. An academic point of view is that IV accurately serves as an expression of the efficient market and, as a result, is an *accurate* forecast of future price movement of the option and its underlying. This theory clearly is false as demonstrated by the difficulty in predicting both underlying and option behavior.

Looking beyond the self-fulling reassurance of the efficient market basis for belief in implied volatility's value, the trader operating in the real world also relies on IV and the broader pricing models of options, in the belief that it enables them to accurately estimate the future. In spite of traders' beliefs (as well as those of the academic community), IV satisfies neither assumption. It is not a predictor of future price movement over the life of the option, but is merely an estimate based on current market prices of options in relation to the value of the underlying security.

This is a profound problem for valuation of options. The widely used IV, with exceptionally complex estimates included in the calculation, is based on flawed assumptions, including the idea that today's stock and option prices are valid starting points for determining future value. In practice, IV varies considerably over an array of strikes and expiration terms, those variables known as the *volatility surface*. This variation makes reliance on current valuation not only inaccurate, but completely backwards. One study concluded, in fact, that changes in stock and option price are the factors that cause changes in volatility; and that there is no evidence that volatility affects future price (in spite of the general belief to the contrary).[4]

The methods for calculating IV, based on current pricing, is assumed to be both reliable and accurate; it is neither. Another study drew this conclusion and noted that IV is far from the valued indicator the market usually assumes it to be:

> In theory, the implied volatility is the market's well-informed prediction of future volatility. In practice, however, the arbitrage trading that is supposed to force option prices into conformance with the market's volatility expectations may be very hard to execute. It will also be less profitable and entail more risk than simple market making that maximizes order flow and earns profits from the bid-ask spread.[5]

The author summarizes the overall flaws in relying on IV as part of the well-known and equally flawed Black-Scholes pricing model (see Chap. 10):

> While the returns volatility of the underlying asset is only one of five parameters in the basic Black-Scholes (BS) option pricing formula, its importance is magnified by the fact that it is the only one that is not directly observable ... Volatility forecasting is vital for derivatives trading, but it remains very much an art rather than a science, particularly among derivatives traders.[6]

Since IV is intended to identify and predict volatility, it is the most popular measurement of risk in options trading. Since volatility is by definition risk, the assumption in the market is that IV is accurate, even though it is not. It contains variables based on current pricing, in spite of the reality that future pricing of stocks and options will be very different (not to mention the variation in moneyness of options and changes in volatility itself as expiration approaches, known as *volatility collapse*). With these problems in mind, notably the unavoidable inaccuracies in IV calculations, traders need a more reliable system for identifying volatility, and to identify probability as it emerges and increases or decreases.

Rejection of IV as an indicator in options trading is questioning of a sacred cow, the absolute belief in an inaccurate estimate of future prices. Just as pro forma financial statements and *forward* P/E are estimates only,

IV bases future price levels on current prices, meaning it does not predict what will occur, only what would occur if today's price were efficient and remained efficient in the future.

So for the trader who wants to better time entry and exit of trades, and who wants to be able to quantify probability as well as risk, what can be used in place of the inaccurate IV formulation? The answer lies in a combination of visual and mathematical signals based on the underlying security and historical volatility, rather than on an attempt to guess the future pricing of the option itself. This assumes, accurately, that option pricing follows the underlying price behavior. This makes historical volatility not only a clearly and accurately defined factor, but one that is reliable for the purpose of estimating future option value.

Chapter 2 includes the formula for calculating historical volatility, including Excel spreadsheet formulas. This allows you to identify relative levels of historical volatility among two or more stocks, for the purpose of determining probability and timing of options trades.

Articulating Risk with Technical Signals

Identifying likely outcomes based on any form of probability analysis is far from simple. In order to attain better than average profits (compared to the larger markets), it often requires taking on more speculative risks, meaning probability has to be modified to make the adjustment, both for profit opportunity and risk exposure. However, traders may easily fall into the trap of believing that probability is somehow "manageable" even when exposure is vastly increased. This is where clearly stated goals and objectives are easily forgotten and put to the side, as speculative behavior replaces conservative behavior. Even those traders who rationalize the decline into speculation as an acceptable aspect of conservative trading, will do well to acknowledge the probability trap, and to recognize that as a form of confirmation bias, this replacement of the original goals and objectives is, in fact, speculative behavior.

This confirmation bias, which may likely have been developed from a simple probability analysis (even when done only in the head and without a specific formula) may mislead the trader, resulting in a puzzled lack of understanding as to why trades are not meeting expectations.

With this analysis in mind, the conclusion is clear: Options traders need to evaluate trades with all many outcomes. This usually encompasses closing to take profits, closing to take losses, accepting or generating exercise, or waiting for expiration. These four outcomes are by no means each likely to contain a 25% probability, even though the average of the four is easily expressed as 25%. In fact, the speculator faces far more daunting likely outcomes than the hedger, who accepts lower returns in exchange for avoiding higher losses—risk transfer.

The nature of options trades, once better understood through basic probability analysis, may be fine-tuned so that the self-defined attributes will translate into practices in line with goals and objectives. Since this is not a statistical book, but one intended to explore options-based pricing and risk functions, the related probability concerns are not addressed here. However, one statistical problem does have to be examined. Options traders face the problem of constantly changing price levels, not only for options premium but for underlying security values as well. The variables include time to expiration, volatility levels, and moneyness of the option. The determination of value varies not only by time but also by moneyness, the relationship between the strike and the current value per share of the underlying.

The range of variables is in constant flux, so standard statistical analysis based on fixed populations is impossible. With options and all of the random variable involved, a more effective method for most calculations is to base valuation and likely outcomes on a probability analysis combining underlying price behavior with other features. Technical indicators designed to identify reversal are likely to manifest in visual signals, so that the determination of option valuation and likely outcomes is more readily identified and put to practical use.

This need for a strong and visual probability mapping system articulates the levels of risk for a particular set of options trades. It mathematically defines the degree of risk to be undertaken to reach your goals and objectives. This is of critical importance in options trading because "risk" is not solely defined by the nature of the strategy. In other words, an uncovered short position is not exclusively high-risk and a covered option is not exclusively low-risk. It fact, the degree of risk to an options strategy is a combination of the specifics of the strategy (long or short, long-term or short-term, at-the-money or not) combined with the equally important aspect of *proximity*.

The proximity between the option's strike and the price level of resistance or support defines risk more than any other attribute, including the specifics of the option itself. This is the case because reversal is most likely to occur at these edges of the trading range. A reversal at mid-range is far less likely than

Fig. 1.1 Tiffany—chart courtesy of StockCharts.com

one at resistance (bearish reversal) or support (bullish reversal). This likelihood is increased when the options signal is accompanied by the underlying price moving through resistance or support, and even more when this dramatic move occurs with price gaps. Given these desirable combinations of price behavior, the option risk is further affected by the existence of exceptionally strong reversal signal and equally strong confirmation.

Considering the mathematical probability of outcomes, a specific strategy is best selected based on the current proximity and price behavior of the underlying, accompanied by reversal signals and confirmation. In this process, the selection and timing are combinations of a thorough understanding of the possible outcomes of a trade, with the purely technical proximity analysis required to fully define risk.

For example, strong reversal and confirmation was observed in the chart of Tiffany (TIF), in which multiple forecasts occurred for bullish reversal, with proximity at support (Fig. 1.1).

The strong bear trend rested at support of $60 per share. As this price level was established, two important reversal signals appeared, a volume spike and a move of momentum into "oversold." The Relative Strength Index (RSI) is a measurement of trend momentum based on comparisons between 14-day averages of upward closings and downward closings. The formula for this indicator is[1]:

$$100 - (100 \div (1 + RS))$$

(RS is the average of upward closings in the past 14 days, divided by the average of downward closings over the past 14 days.)

RSI is a strong reversal signal because moves into oversold or overbought are rare. In the Tiffany chart, this move below the index value of 30 is the only occurrence in 6 months. So RSI confirms the location of the volume spike as a high-probability reversal point. However, even more confirmation is found in two additional signals found two to three weeks later. A second set of volume spikes serves as a reminder of the likely adjustment in price behavior. The conclusive form of confirmation is in the three-session candlestick, a bullish morning star. Note that this occurred at support, with the middle session moving below that level but closing at the support price.

The question worth addressing when patterns such as this emerge is: What options trade is appropriate given the technical signals? The mathematical strength of RSI provides a strong indication of the coming reversal, but for the options trader, the question is more complex. Many traders favor a particular strategy or a short list of favored strategies, but avoid other strategies perceived as high-risk. However, given the strength of the multiple signals on the chart, the true risk of a particular trade in these circumstances is vastly different than the same trade with weaker proximity. This means that strategies perceived as "low risk" or "high risk" might not always be properly defined. For options traders preferring specific strategies over others, this raises the point that "risk" by definition has to be a variable given the circumstances: strength of signals and confirmation, proximity of signal to resistance or support and the attributes of the trade itself. However, these attributes are not the sole defining factors of risk, but only part of the larger risk equation.

In establishing trading goals and objectives, this variation in risk itself may serve as an expanded form of analysis. By combining the mathematical probability of success in a trade, with the equally essential math of momentum, and then observing other price and volume signals, the conclusion is that risk consists of much more than the attributes of specific trades.

---

[1] Fortunately for traders relying on chart indicators and momentum oscillators, indicators such as RSI are calculated automatically and added to the chart. The formula is included here for the purpose of demonstrating the components of the indicator, in order to enhance understanding of exactly what it reveals.

Probability in an Uncertain World

Options traders cannot rely on normal distribution used in statistics, because the population of possible outcomes is not fixed. This reality further supports the contention that implied volatility cannot be used to reliably estimate future option valuation.

The normal statistical methods for identifying a range of possible outcomes (distribution) is based on a fixed population. For example, voting patterns among registered voters involve a finite number of choices (votes) among a fixed and well-understood group of individuals (registered voters). This fixed nature of "normal distribution" is reassuring in statistical analysis because there is a limited range of possible outcomes—selection of one candidate or another, a decision to not vote at all, or the wild card of a write-in decision.

The finite likely outcomes, when placed on a bell curve, illustrate how standard deviation works. Standard deviation is identified in formulas by the Greek letter sigma, σ. Normal distribution describes the most likely outcomes, represented on the bell curve. An example is shown in Fig. 1.2.

This reveals the application of standard deviation. With normal distribution (for example, in analysis of voting among registered voters), 68% of outcomes will fall within one standard deviation, and 95% within two standard deviations. This is a reliable statistical fact assuming normal distribution, but how does this help the options trader? Given the variances in

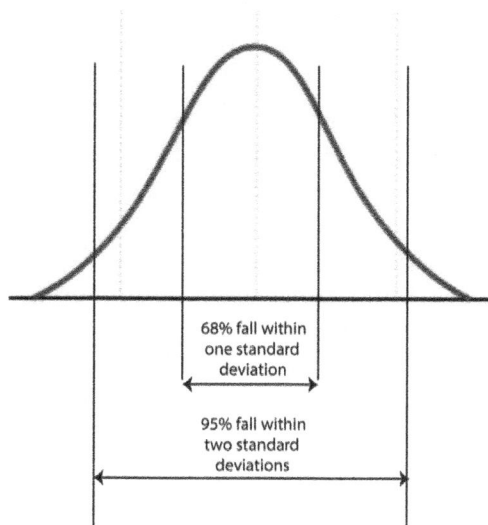

Fig. 1.2 The bell curve—prepared by the author

underlying stock prices, moneyness of options, and volatility, options markets never have normal distribution, making probability analysis elusive. However, there are solutions.

To develop a reliable probability analysis for options trading, its purpose has to be understood as a first step. Because options treading is a matter of skillful timing for both entry and exit, depending on simple additive probability calculations is not accurate. An example of an additive calculation is the previously demonstrated roll of either one die or two dice. An additive probability for four rolls of one die or 24 rolls of two dice was 67% for any specific outcome. However, by calculating likely failure outcomes and then subtracting from '1' yields a multiplicative result that is far more accurate. To understand how to set up a workable system to overcome the elusive nature of options trading, standard deviation is useful as a starting point. In fact, it solves the problem by identifying how to make those random variables predictable and highly visible. If this visual form of probability can be used to then yield actionable reversal signals, it solves the options challenge.

Standard deviation often is not well understood. As a statistical calculation, it applies in specific ways, and this may also be useful in analysis of the timing for options trades. An example of how this is calculated: An examination of 50 men and 50 women reveals that the average life span is 76 (men) or 81 (women). Within this group, the standard deviation is approximately 3 years. So male lifespan is between 73 and 79; and female lifespan is between 78 and 84. To apply this, 68% are likely to fall within one standard deviation, so that 34 men and 34 women will experience lifespan within the three-year ranges. Applying the bell curve with two standard deviations, 95% of both groups will experience lifespan of two standard deviations, or a range of 6 years from the average—70 to 82 years for men and 75–87 years for women.

Because standard deviation is a strong statistical averaging method, when applied to the creation of a reliable probability map based on stock price movement, it provides two benefits to options traders. First, it represents a mathematical range of likely price behavior. Second, the mapping of probability itself generates strong reversal signals that can be used to identify timing of entry and exit in an options trade.

To calculate standard deviation, six steps are involved:

1. Determine the average (A). The time period involved defines the starting point of the matrix. For example, using 10 consecutive periods means the values are added together and then divided by 10:

$$(N_1 \ldots N_{10}) \div 10 = A$$

2. Calculate each period's deviation (D). This is the difference between each value and the average for the entire field:

$$N_X - A = D$$

3. Calculate the square (S) for each session's deviation. Each deviation is multiplied by itself:

$$D^2 = S$$

4. Add the squared values (SV). In the case of 10 periods, the 10 squared values are added together.

$$S_1 + \ldots S_{10} = SV$$

5. Find the average. Divide the sum of the squared values by the number of periods.

$$SV \div 10 = A$$

6. Calculate the square root of the average. This is one standard deviation of the field.

$$\sqrt{A}$$

Taken in these six steps, the formula is not difficult to follow. A summarized and more complex version of the formula for standard deviation is shown in Fig. 1.3.

This formula states that standard deviation σ equals the square root of: addition of values (**N**) from 1 to the total Σ, multiplied by the square of the individual values $\chi 1$, minus the average μ.

This formula can also be calculated on an Escel spreadsheet, by entering the following:

standard deviation

$$\sigma = \sqrt{\frac{1}{N}\sum_{i=1}^{N}(x_i - \mu)^2}$$

σ standard deviation
N   addition of values
Σ range of values from 1 to n
$\chi 1$ individual values
μ average

Fig. 1.3 Standard deviation—prepared by the author

Column A—enter each value in the field; for example, if 10 values are involved, this will fill up cells A1 through A10.

Column B—Find the average of the values in column A, and enter the average in each row. The Excel entry for cell B1 is:

$$= SUM(A1:A10)/10$$

Copy the result into each of the remaining cells in column B.

Column C—Find the deviation between each value and the deviation. The Excel entry for cell C1 is:

$$= SUM(A1 - B1)/10$$

Copy the result and paste into the remaining cells in Column C; the result is calculated automatically for each deviation.

Column D—square each value in Column C. The Excel value for cell D1 is:

$$= SUM(C1*C1)$$

Copy the result and paste into the remaining cells in Column D.

Column E—Add the squared values and enter in the last row of Column E. Based on the example, the Excel formula for cell E10 is:

$$= SUM(D1:D10)$$

Column F—find the average of the totaled values. The Excel formula for cell F10 is:

$$= SUM(E10/10)$$

Column G—find the square root of Column F. The formula for cell G10 is:

$$= SQRT(F10)$$

The result is one standard deviation.

This set of formulas in an Excel spreadsheet yield the results based on the values shown in column 1, on Table 1.1.

Normal distribution, represented visually by the bell curve, is easily used to identify where most outcomes will fall for many situations. However, when it comes to options trading, normal distribution cannot be used because the range of possible outcomes is constantly changing. Unlike the sample of registered voters deciding among a limited number of candidates, the outcomes are clearly finite. The continuous random variables seen on stock charts and among option premium outcomes translates to an inability to expect normal distribution. Even one element alone, the stock price,

Table 1.1 Excel formula, calculation of standard deviation—prepared by the author

(A) Value	(B) Average SUM(A1:A10)/10	(C) Deviation = SUM(A1-B1)/10	(D) Square of each deviation = SUM(C1*C1)	(E) Add the squared values = SUM(D1:D10)	(F) Find the average of column E = SUM(E10/10)	(G) Find the square root of Column F = SQRT(F10)
105.58	109.77	−0.42	0.176			
107.05	109.77	−0.27	0.074			
110.44	109.77	0.07	0004			
109.88	109.77	0.01	0.000			
110.51	109.77	0.07	0.005			
110.32	109.77	0.05	0.003			
111.32	109.77	0.16	0.024			
110.16	109.77	0.04	0.002			
110.57	109.77	0.08	0.006			
111.88	109.77	0.21	0.045	0.339	0.034	0.18

may experience a range of possible closing price between zero and infinity (in spite of more *likely* ranges of outcome). However, with the use of resistance and support as a starting point in constructing a probability matrix, the certainty of a mathematical probability formula (standard deviation) enables the orderly progress toward a visual matrix that defines likely results and also helps identify reversal signals.

This is where the mathematical construction of what may be termed a probability matrix is beneficial in timing of trades. Unlike the well-known reversals found in Western technical analysis, candlesticks, volume, momentum and moving averages, an effective probability matrix produces reversal signals combining the price pattern with mathematical trends. However, the range of possible outcomes for any stock or option value (known as the probability density function) remains elusive given the problem of continuous random variables. The solution to this challenge is in translating the mathematical calculation of standard deviation into a charting tool useful in identifying price movement as well as reversal with either confirmation or divergence.

The well-known bell curve reveals a clustering of outcomes within either one or two standard deviations. In the markets, however, this does not occur due to the dimensions of variables. What would be rare in most statistical outcomes, the *fat tail*—outcomes well outside of normal distribution—is common in the stock and option markets:

> … if returns were "normal," statisticians would expect the S&P 500 to move up or down by 3.5% or more only once every 10,000 years … In contrast, we've experienced 118 such occurrences since 1950–nearly half of them in the past two years. Over the past two years, the average daily move up or down has been 1.3%. With that as a baseline, a normal distribution would see a fat-tailed move of 6.4% once every 100 years. In fact, it has already happened 11 times in the past two years.[8]

Given the problem in the stock and options markets of continuous random variables, how can a trader determine not only valuation, but also timing for trade entry and exit? It might seem impossible, but the math itself points to a solution. The probability matrix is not only a visual summary of probability, but also a specific signaling indicator of its own. This indicator, the *Bollinger Bands* named for its developer John Bollinger, reflects the nature of stock price behavior in a manner enabling the prediction of price movement. This in turn helps predict the best timing for options entry and exit. Thus, a signal developed from the well-known statistical calculation of standard deviation transforms the *random* into the *likely*.

Because Bollinger Bands is automatically calculated via free online charting services, it is an easy indicator to employ. It converts the math into a convenient and visual reflection of the probability range. For options traders, this makes the technical signal a desirable alternative to the less certain calculations of probability so often used in options trading. Since the bands reflect averages of the underlying price and its standard deviation as explained in the following section, it offers a math-based tool for the timing of options trades based on stock price behavior rather than on the more elusive implied volatility of options.

Bollinger Bands to Create a Probability Matrix

The powerful set of *bands* that collectively make up Bollinger Bands (BB) creates a visual nature of price behavior derived from price averages and standard deviation. The middle band is a 20-period simple moving average of price. The upper band represents the middle band plus two standard deviations; and the lower band is the middle band minus two standard deviations.

To describe exactly what BB accomplishes, its original developer, John Bollinger explained:

> Bollinger Bands are bands drawn in and around the price structure on a chart. Their purpose is to provide relative definitions of high and low; prices near the upper band are high, prices near the lower band are low.[9]

This simple explanation is both accurate and elegant. However, while BB provides an initial visual reflection of price behavior and high or low status, it can be used for much more. The pattern of the three bands creates a probability matrix because it represents the likely range of price extremes you would expect to see in a well-behaved trading pattern. So when price misbehaves based on this matrix, meaning when it moves about the upper band or below the lower band, a change is signaled—this means either that the current range is likely to change as price moves to new levels, or that the breakout will lead to reversal.

The range from upper to lower bands accomplishes even more. The levels of activity become a form of dynamic resistance and support. Most traders are accustomed to seeing straight lines representing these trading range levels, but BB takes this a step farther and builds a dynamic representation of these price levels. The tendency is for BB's upper band to closely track resistance during uptrends, and for the lower band to closely track support during downtrends. The actual band width also visually represents volatility, making changes in volatility visible as well.

The method of calculation for the three bands solves the problem of statistical inaccuracy. With continuous random variables, the bell curve cannot be applied to ever-changing stock prices nor to option valuation. BB resolves this challenge at least for stock price behavior. In turn, this points to the timing for options entry and exit, all based on standard deviation for the single factor of a stock's closing price for each session. The value in this single indicator is multi-faceted as it identifies changing price volatility (band width expanding or contracting), breakouts (trading above upper band or below lower band) and reversal signals. The reversals make BB especially useful during periods of consolidation. With prices range-bound, many traders believe that no signaling decide is useful since there is no dynamic trend to reverse. This belief ignores one reality: A period of consolidation is itself a trend, moving sideways rather than up or down.

When consolidation is in effect, many of the signals BB produces are more valuable than at other times. For traders acting on reversals, it makes sense to redefine *reversal* itself to indicate activity in a trend and not just in price. In other words, if consolidation is a third type of trend (in addition to bull or bear), then a move out of consolidation is a reversal of that trend. This means that a reversal signal may also be applicable. Analysis of price-based signals as well as non-price reversals indicates that this is a reasonable assumption.

For example, Fig. 1.4 reveals how the mathematical aspects of Bollinger Bands set up the visual breakout from consolidation, and how reversals mark and confirm the breakout. The stock had been trading in consolidation between $58 and $54 from March 7 through April 30. At that point, several events occurred within three weeks, marking a bullish reversal. First was the bullish doji star, followed by the piercing lines with a 2-day volume spike. After this, a price gap occurred leading to yet another bullish signal, the harami cross.

Even with all of these bullish signs, the probability matrix established by BB reveals further confirmation. When prices moved below the lower band (and below support) during the month of May, the fast reversals signaled a failure by the bears to move the price lower. Even though price later moved above the upper bands, no reversal occurred, indicating a likely successful breakout to a higher trading range.

The BB signals were also found in volatility changes. In March, the band width was quite large, about 10 points. This narrowed to only four points at

Fig. 1.4 Alamo group—chart courtesy of StockCharts.com

the point of the breakout, another confirmation that a change was likely to occur, As the breakout began, volatility increased once again, visually seen in the band width as it grew to 11 points.

This chart is a good example of how BB transforms the standard deviation of price into a visual forecasting mechanism of price breakout. The candlestick signals and price gaps are recognizable without BB added to the chart, but BB provides strong confirmation of the reversal in the form of trading outside of the bands, and changes in volatility in the form of band width.

The visual representation of the math behind BB makes the point that the very patterns created in this way generate technical signals based on price averages. As a confirmation signal, BB patterns work in both bull and bear trends. In all types of trends, two particular BB patterns signal reversal in a very strong manner. These are the M top and the W bottom. These are similar to other technical signals, notably double tops or bottoms and the popular head and shoulders. However, the M and W patterns are located within the movement of the BB range.

The M top is a bearish signal, so named because it forms a letter M. Given the tendency of some stock prices to whipsaw back and forth, it would be simple to interpret the M top as a W bottom by drawing the lines in different places. However, one attribute defines the completion of the M

1 Trading Goals and Objectives   21

Fig. 1.5 Matson—chart courtesy of StockCharts.com

top: the last leg has to move and close below the lower band. For example, Fig. 1.5, the chart of Matson (MATX) shows an M top from mid-March to early May. This forecasts a bearish reversal out of consolidation. It is a legitimate signal because the last leg of the M closes below the lower band. Confirming this signal was the volume spike occurring with completion of the Bollinger signal, as well as a strong downward price gap.

The W bottom is the opposite, a bullish signal occurring within the range of the BB and with the last leg closing above the upper band. Like the M top, the W bottom presents a visual summary of BB's mathematical averages. An example is seen in the chart of IBM (IBM) the W bottom is marked, validated by the closing of price above the upper band on the last leg of the W (Fig. 1.6).

The pattern is confirmed by the upside tasuki gap, a bullish continuation indicator. In this situation, the price pattern continued upward as expected, until April when it settled into a new consolidation trend between $145 and $150 per share.

The mathematical nature of BB is highly reliable not only because it yields signals on its own, but also because it sets up a trading range with a high degree of reliability. Movement of price outside of the zone between upper and lower bands should be viewed as a signal moving away from the two

Fig. 1.6 IBM—chart courtesy of StockCharts.com

standard deviations of the price average, which is significant. In establishing the risk aspects of goals and objectives, BB is one of the best indicators for ensuring that emerging changes in price behavior are observed as early as possible. In addition, the exceptionally strong mathematical attributes of BB present a reasonable alternative to popularly used probability formulas. In fact, up to 90% of all price activity is statistically likely to close within the range between upper and lower bands.[10]

Beyond the M top and W bottom, BB creates even more high-value signals. Among these is the *Bollinger squeeze*, an extreme narrowing on daily breadth of trading. The squeeze takes on forecasting significance when it moves toward the upper band (bullish) or the lower band (bearish). The theory behind this pattern is that narrowing range is followed by broadening range and, often, by a breakout in the indicated direction. In consolidation trends, the squeeze is likely to signal a coming end to the trend and beginning of a new and potentially volatile bullish or bearish trend.

An example of the Bollinger squeeze was seen on the chart of J.M. Smucker (SJM), in Fig. 1.7. This contained multiple signals confirming the Bollinger squeeze and anticipating the end of consolidation.

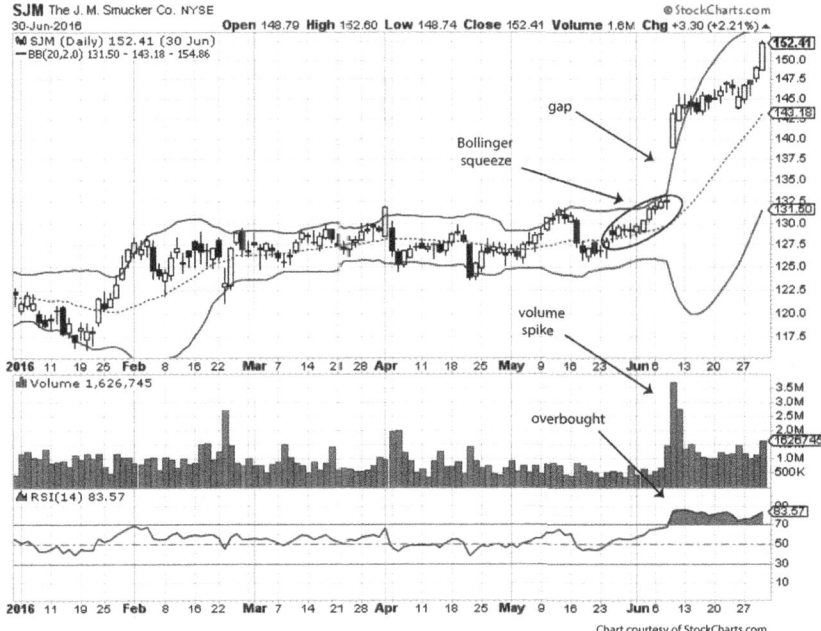

Fig. 1.7 J.M. Smucker—chart courtesy of StockCharts.com

Consolidation between $125 and $130 had been in effect since March, and lasted until the end of May. The highlighted squeeze showed price trending toward the upper band as daily breadth narrowed. As price broke through resistance with a large upside gap, it was confirmed by a volume spike and movement of momentum into overbought. This collective group of signals was exceptional strong. As expected, narrow breadth was followed by higher breadth and increased volatility. The BB bandwidth moved from the consolidation breadth of five points to more than 20 points after the breakout.

The examples of BB patterns and their signal value demonstrates that a mathematical function may be made visible and actionable, enabling traders to base entry and exit on forecasts in many forms. This strengthens goals and objectives based on risk management. As the patterns develop and change is anticipated as shown in BB and confirming signals, the prevailing objectives should be used as guidelines for the timing of new trade decisions. This defines the difference between conservative portfolio management with options, and speculation.

Speculation Versus Hedging

The goals and objectives for conservative options trading are well defined when based on formulations such as Bollinger Bands. The mathematical rigidity of the two standard deviations makes BB an excellent form of probability matrix. This ensures that goals and objectives for conservative options trading can be tracked and estimated reliably. So in a hedging and income-generating program objective, the path involves application of technical signals included BB, candlesticks, volume signals and momentum. However, in a program of speculation where risk tolerance is much higher, the mathematical emphasis tends to be on calculations of volatility and probability. The usual speculative options strategy is based on timing of trades based on ever-changing levels, especially of volatility.

The speculator enters into a trade not with long-term hedging in mind, but primarily to generate quick profits as soon as possible. As a consequence, the types of trading activity tends to be exceptionally high-risk. The speculator makes a judgment call, accepting exceptionally high risks in the hope of generating exceptionally high profits. This also means that speculation generally ignores fundamental factors except to the extent that they aid the timing of the speculative trade. For example, timing of a trade anticipating an upcoming earnings report and a big price move, is one example of speculative use of a fundamental (earnings).

If an options trader accepts the risks associated with speculation and embraces the risk profile this requires, then it is an informed choice. Of much greater concern is the relationship between a conservative set of goals and objectives that are violated through specific high-risk and speculative options trades. This may occur for several reasons:

1. *Identification of an unusual opportunity.* The conservative investor might see a price pattern or read news concerning a company, and become convinced that a short-term profit opportunity is possible. This ignores the informational efficiency of the market, the observed fact that all underlying prices are immediately discounted for known news about a company. This efficient market hypothesis (EMH) often is misunderstood in the assumption that it means the market is efficient overall. However, EMH does not mean this at all. It only identifies the efficiency of how information affects security prices. The disturbing fact about this is that all "news" is discounted with the same efficiency, including true news as well as false reports, rumors, gossip and exaggeration.

The tendency to act on what appears to be an unusual opportunity may violate the carefully defined goals and objectives of a conservative trader. So acting on what appears to be facts everyone else has overlooked, tends to mislead and distort. As a consequence, ill-timed trades or selection of high-risk trades may lead to unexpected losses. One tendency is to select a high-risk trade due to the apparent unusual opportunity, even though it violates the risk tolerance defined within goals and objectives.

2. *The desire to recapture losses on a previous trade.* An especially destructive behavior is to try and recapture the loss of a previous trade by increasing the risk level of a future trade. This is self-destructive for anyone if the selection of a strategy violates the risk tolerance. This tendency—to increase risks to recapture losses—leads to more reckless behavior in the future. If the attempt fails, an even greater risk might be entered to recapture a greater loss. If the higher risk succeeds, it might convince the trader that speculative risks are worth taking—again, violating the carefully structured risk tolerance within goals and objectives.

The more rational way to treat losses is to put them behind you and move forward, without changing the risk levels of future trades. A mature options trading philosophy recognizes that losses are going to occur. This demands a sensible mindset and the ability to learn from mistakes but also to accept them. No single strategy presents 100% certainty. Using strong technical signals and confirmation improves the percentages of well-timed entry and exit, but as part of this program, losses are a fact of life.

3. *A failure to recognize the risk level of a specific strategy.* One of the most common errors made by options traders, including even the most experienced, is to enter trades containing risks not recognized at the time. For example, a vertical spread mitigates short-side risk and limits it to the strike distance between a long and short position. However, if the long position becomes profitable and it closed, it leaves an uncovered short position open. This could easily violate the risk tolerance levels for that trader, and expose the position to danger of exercise. Another example is the most popular of all options strategies, the covered call. Many traders consider this a risk-free sure thing type of trade. However, if the underlying price declines beyond net basis (price of shares minus premium received for the short call), it sets up a paper loss. Recovery is possible only through subsequent offsetting strategies or by waiting out price in the hope it will rebound.

The solution is to time covered calls to emphasize slightly OTM contracts with expiration in one to two weeks, and to avoid entering positions beyond that. Another solution is to examine the alternative of an uncovered put. This has the same market risk as the covered call, but with the greater flexibility of being able to roll forward to a different strike. The covered call writer has to be aware of exercise risk with strikes below original cost per share, but with uncovered puts, it does not matter which strike is utilized. The sensible approach to strategic selection and examination of all possible outcomes helps avoid the mistake of not identifying some of the risks.

In spite of high risks, speculation with options remains highly popular. The overall trend since 1973 when standardized options were first introduced has been for options trading to focus increasingly on portfolio management through hedging. In the past, however, improved efficiency in the markets led to greater speculation in spite of evidence that the risks were too great for many. In 1900, the estimate number of speculators was less than 5 million. Three decades later, the number had grown to 26 million.[11]

With the introduction of the Internet, volume for all types of trading and investing has increased once again. In 1973, a very small number of options contracts were traded, on a limited number of stocks. Only calls were traded, with puts not introduced until later. Figure 1.8 reveals the curve of growth in options trading over many years.

The volume combines both speculation and hedging activity. The trend is for a growing number of professional investors and portfolio managers to use options as hedging devices to reduce and eliminate market risk in equity portfolios. However, speculation remains a common practice in the options market. Some speculators rely on formulas to time trades, and others act in a more intuitive manner, replying on their perception—like all speculators—that their market insights are enough to skillfully time trades. This is a false assumption, and a majority of speculators lose more than they gain.

The major differences between investing or trading on one hand, and speculating on the other, focus mainly on risk. A "high" risk is associated with speculation, but for many this means risk not of market decline but of exposure to market risk. A decline is not the same as a risk. A market decline tends to be cyclical, trend-related, or the outcome of retracement. A risk, specifically market risk, usually means that a security is overpriced when bought and that the subsequent loss will not be recovered in the short term, if at all. Those who invested (or, speculated) in the dot.com or tech bubbles two decades ago, or in the housing market before 2008, understand this distinction all too well.

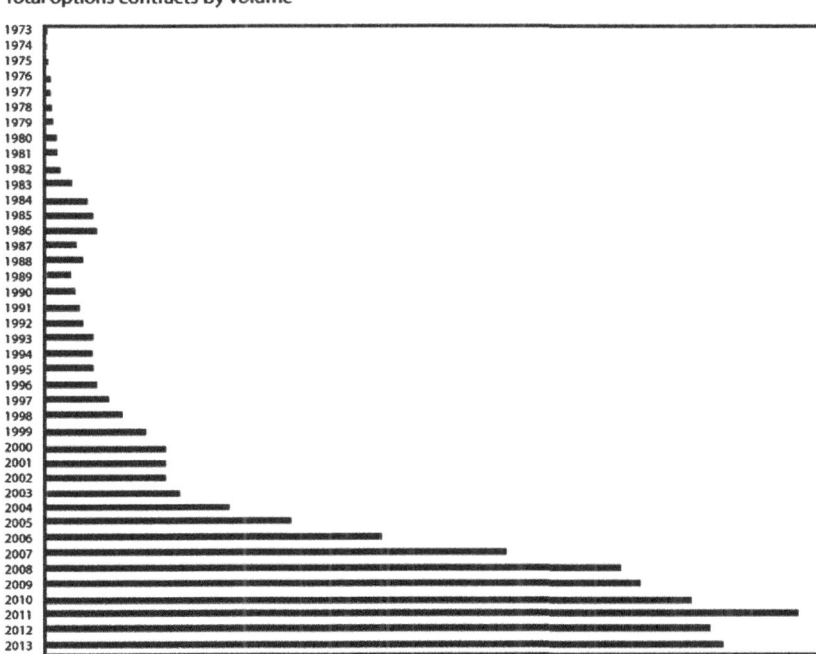

Fig. 1.8 Total options contracts by volume—prepared by the author

Speculation, however, provides a lot of liquidity and order in the market. For many who prefer speculation over investing or hedging, the availability of options for speculative goals and objectives certainly makes it easier that it was before standardized options trading was available. Many conservative market analysts and investors, both institutional and retain, condemn speculation as unethical, greedy, or immoral, based on a motive of greed through exploitation of the more conservative investor or trader. This opinion deserves to be questioned, however, because speculation, with its greater than average risks, benefits the market and the economy.

The supply and demand cycle often becomes chaotic, and at such times speculators tend to even out the distorted side of the economic cycles. Because speculators identify and exploit scarcity, stock prices and option premium are forced upward, increasing the participation phase of trends and leading to even higher profits.

By placing money at risk, speculators also add liquidity to a market that could otherwise suffer from illiquidity, making it more difficult for conservative investors to take part. This adds to market efficiency, which would be erratic and at times impossible without speculators. For risk-averse investors and traders, speculation affects prices and trends and this, in turn, encourages others to also risk capital. Speculation creates value for shares of stock and for options through *price discovery*, the result of trading among buyers and sellers. This is essential for options pricing, and without speculation this would be difficult and perhaps inaccurate at times.

On the other side of these arguments, speculation also sets up pricing bubbles. The 2008–2009 housing market bubble was the consequence of speculation on the part of debt obligation brokers, real estate speculators, and even new homeowners. Regulation aimed at curbing speculation over time has been ineffective for the most part. In the U.S., one of the best-known of these was the Glass-Steagall Act of 1933, part of the post-market crash legislation that created the Securities and Exchange Commission and related laws. The law contained a range of provisions, but the most important for the investment market was a requirement to separate commercial and investment banking. More recent legislation aimed at speculators included the Sarbanes-Oxley Act and the Dodd-Frank Wall Street Reform and Consumer Protection Act. These new laws and regulations may improve disclosures, eliminate conflicts of interest or inaccurate financial reporting, and mandate stricter limits on corporate executives. But ultimately, speculators will find ways to place capital at risk in the desire for ever higher and faster profits. If the speculator's goals and objectives are specifically designed to pursue higher profits through taking higher risks, regulation will not stop them.

For the same reasons, conservative investors who define and practice safer, lower-risk goals and objectives in their own stock and option portfolios can never rely on ever more stringent regulations to protect them from experiencing losses in the market.

Chapter Summary:

- investing objectives are unavoidably factors of risk levels
- implied volatility is an estimate only and a poor indicator for timing trades
- price leads volatility; a flawed belief is that volatility creates price movement
- strike proximity to resistance or support determines the strength of a trade
- a probability matrix creates a dramatic and visual trade timing system
- Bollinger Bands is a powerful visual signal for trade timing.

Notes

1. Hirshleifer, J. (1977, September). The theory of speculation under alternative regimes of markets. The Journal of Finance, 32, 975–999.
2. Tirole, J. (1982, September). On the possibility of speculation under rational expectations. Econometrica,50 (5), 1163–1181.
3. Eves, Howard (1990). An Introduction to the History of Mathematics. Fort Worth TX: Saunders College Publishing, p. 354.
4. Bouchard, Jean-Philippe & Marc Potters (2009). Theory of Financial Risk and Derivative Pricing: From Statistical Physics to Risk Management (2nd ed.). Cambridge UK: Cambridge University Press. p. 252.
5. Figlewski, Stephen. (2004). Forecasting Volatility, New York University Stern School of Business, Preface.
6. Figlewski, Op. Cit., p. 1.
7. Fisher, Gregg S. (2009, October 14). How to protect investments from cataclysmic 'fat tails'. Forbes.
8. Bollinger, John (2001). Bollinger on Bollinger Bands. New York: McGraw-Hill, pp. xxi–xxii.
9. Grimes, Adam. (2012). The Art & Science of Technical Analysis: Market Structure, Price Action & Trading Strategies. Hoboken NJ: John Wiley & Sons, pp. 196–198.
10. Stäheli, Urs (2013). Spectacular Speculation: Thrills, the Economy, and Popular Discourse. Stanford, CA Stanford University Press, p. 4.

2

The Role of Fundamental and Technical Analysis

Chapter Objectives:

- study the direct relationship between fundamental and historical volatility
- develop an efficient method for calculating historical volatility
- compare the advantages of historical volatility to the problems of implied volatility
- reveal the correlation between fundamental volatility and stock price behavior
- analyze the comparisons between fundamental volatility and options risk
- rate proximity with a point system to appreciate the probability of success.

Fundamental volatility (defined as trends in financial outcomes for a company) directly and at times significantly impacts a stock's historical volatility and, as a result, options status as well.

This claim might surprise many options traders, who tend to dismiss fundamental analysis as backward-looking and of no value in (a) selecting specific trading strategies; (b) identifying volatility or risk levels; and (c) establishing realistic profit expectations. This chapter reveals how and why fundamental volatility is part of the equation for determining option volatility and the timing of trades.

Analyzing the Impact of Fundamental Volatility

The rejection of the fundamentals among options insiders overlooks a significant source of intelligence for the timing of trades and an opportunity to expand the overall understanding of market risk. A mathematical analysis of key fundamental signals reveals a correlation between long-tern fundamental volatility and the options market. Thus, the selection of a company and its stock as a vehicle for options trading can be shown to rely on fundamental analysis as a starting point and, ultimately, as an influence on volatility in both stock and option prices. If a trader is intent on building a portfolio of value investments and then hedging market risk with options, awareness of fundamental trends identifies volatility tendencies within the options market.

In the context of pricing for both stocks and options, "fundamental volatility" refers to a tendency in a company's financial statements to display varying levels of predictability. Thus, the financial trends observed in revenue, earnings, long-term debt, dividends, and P/E ratio are examples of fundamentals that define levels of financial stability and predictability.

However, in literature analyzing markets, the term "fundamental volatility" may also refer to *economic* fundamentals including GDP, consumption and other measurements. These macroeconomic variables are not associated directly with the correlation between a company's reported profitability or cash management, however. In the analysis that follows, "fundamental volatility" is a description of financial trends over time, with a study of how these trends relate to stock trends as well as option pricing. Even though the economic and financial definitions of volatility are dissimilar, the conclusions are worthwhile: The relationship between a company's financial trends (fundamentals) and stock price trends (technicals) is correlated, and this presents a crucial element in selection and timing of stock investments and option trades.

However, a problem persists in the methods used by traders within the options market. These traders tend to rely solely on technical indicators associated with stock price behavior (technical analysis) and estimates of future option value (implied volatility), while ignoring and discounting the value of fundamental volatility of financial trends and its effect on historical volatility of stock prices. This discounting of fundamental volatility occurs in spite of the direct correlation between price behavior and fundamental volatility. As one study discovered, accounting trends

… exhibit incremental predictive power with respect to future option returns above and beyond what is captured by implied and historical stock volatility, suggesting that the options market does not fully incorporate fundamental information into option prices.[1]

A related problem to this discounting features in the options market is a tendency for reduced levels of pricing efficiency, compared to the more immediate discounting within the stock market. While stock price behavior is assumed to the informationally efficient (meaning information is taken into price and discounted immediately), the same is not necessarily true in the options market. Informational efficiency means that stock prices react immediately to information (both true and false); however, a distinction has to be made between the long-term impact of fundamental volatility and the short-term impact of technical price volatility of stocks. To the extent that these factors affect option valuation, the fundamentals tend to define equity value over the long term, and this in turn serves as a starting point in selecting equity investments for inclusion in an options-based hedging program.

The efficiency or lack of efficiency in options pricing behavior is apparent in comparisons between stock investment versus options leverage; and differences in volume levels between the two:

> On one hand, the leveraged nature of option contracts attracts sophisticated investors who wish to exploit public and private information. On the other hand, several institutional features of the options market make it less efficient than the stock market. For example, an option contract based on a firm's stock typically has considerably lower trading volume than the stock itself.[2]

Yet another factor in the exchange between fundamental volatility and options pricing is related to earnings surprises and changes in management guidance. The impact of both positive and negative earnings surprises is immediate and easily observed in stock charts. For example, in late February, JC Penney (JCP) experienced a positive surprise of 72.3% above expectations. The price gapped higher and was strongly confirmed by a volume spike and momentum moving into the "overbought" index range (Fig. 2.1).

The immediate impact of earnings as a fundamental indicator on the technical behavior of price is one example of how the fundamentals affect the technicals, in this case immediately. The clear reversal signals mark the logical point for entering trades, and options traders relying on chart analysis improve their timing by observing these reversal and confirmation signals. However, the analytical aspects of stock price behavior are only the first

Fig. 2.1 JC Penney—chart courtesy of StockCharts.com

step in relating the same level of news to options trading and, specifically, in how that shows up in volatility. In fact, many studies have concluded that closely related to earnings, management's guidance concerning future revenue and earnings forecasts also has a direct effect on the volatility of options, which should come as no surprise. Options implied volatility (IV) as well as a stock's historical volatility (HV) is related directly to news released about the underlying corporation. One study reported that "the implied volatility values increase after managerial forecasts, particularly when the forecast conveys bad news."[3]

Earnings surprises and changes in guidance forecasts are immediately observable, but these are not the only fundamentals that can be traced to technical price behavior. For example, announcements concerning dividends (declared, raised, lowered, or skipped) are fundamentals directly impacting stock price volatility and as a result, options volatility. A corporation's "decision to pay a dividend signals a commitment to maintain that dividend, implying a level of stability in the firm's operations. Thus, managers can use a dividend to signal lower fundamental volatility."[4]

This rationale may overstate the impact of fundamental influence, even in regard to dividend news. The point remains, however, that such an announcement has implications for fundamental volatility levels in the future. This change must be expected to also be reflected in historical volatility of the stock.

The relationship between fundamental news and trends, and the consequential options volatility, is not apparent to all, and is taken for granted by many options traders. However, it is also possible to demonstrate through analysis of fundamental volatility over many years that a very direct correlation exists between the fundamental and the technical. The question remains, once this correlation is observed, whether to rely on options implied volatility, or on stock historical volatility. The flaws of reliance on implied volatility were examined in detail in Chap. 1. With these flaws in mind, reliance on historical volatility provides a more reliable and more accurate measurement of options price risk and opportunity.

Calculating Historical Volatility

Historical volatility is based on daily stock prices at closing. The calculation reveals the standard deviation of net returns from one day to the next, and annualized over the full year.

Options traders may equate fundamental volatility with stock price historical volatility with a high degree of accuracy. An analysis of long-term trend in each proves this point. This direct correlation makes a compelling case for identifying levels of historical volatility to better understand option volatility as a defining factor in risk for options strategies.

To demonstrate how fundamental volatility of a company and historical volatility of stock prices are closely related (and as a result, also affect an option's implied volatility), consider the following example:

Historical volatility over 10 days: A stock has traded over the last 10 days with the following closing prices: $105.58, 107.05, 110.44, 109.88, 110.51, 110.32, 111.32, 110.16, 110.57, and 111.88. To calculate historical volatility, use the following formula in an Excel spreadsheet:

Column A—enter each day's closing price (in the preceding example, 10 consecutive trading days were used).

Column B—calculate the daily net change (divide each day by the prior day, and subtract 1.

Column C—multiply Column B by 100. Formula is = **SUM(C1 * 100)** Copy and paste to other cells

Table 2.1 Excel formula, calculation of historical volatility—prepared by the author

(A) Closing price each day	(B) Return (divide each day by prior day) = SUM(A2/A1) − 1	(C) Multiply column B by 100 = SUM(C1 * 100)	(D) Standard deviation = STDEV(C1:C10)	(E) Annualize = historical volatility = SQRT(252) * D10
105.58				
107.05	0.0139	1.39		
110.44	0.0317	3.17		
109.88	−0.0051	−0.51		
110.51	0.0057	0.57		
110.32	−0.0017	−0.17		
111.32	0.0091	0.91		
110.16	−0.0104	−1.04		
110.57	0.0037	0.37		
111.88	0.0118	1.18	1.30	20.68%

Column D, last 10—calculate the standard deviation of Column C. The formula to enter into D10 is: = **STDEV(C1 : C10)**

Column E, row 10—annualize the standard deviation based on average trading days per year of 252. This is the square root of standard deviations. The formula for cell E10 is:= **SQRT(252) * D10**[1]

This set of calculations is summarized in Table 2.1.

In this example, historical volatility is determined to be 20.68%. As a relative value, compare this to volatility at different times over similar periods, or to other stock price movement to judge the market risk of stock (and as a result, to also time entry and exit of options trades).

The value to options traders in the use of standard deviation to quantify historical volatility becomes of greatest value when the outcome is extreme: "The standard deviation is a simple but useful measure of volatility because it summarises the probability of seeing extreme values of return. When the sample standard deviation is large, the chance of a large positive or negative return is large."[5]

The Problem with Implied Volatility

To compare historical volatility of a stock to implied volatility of an option is a comparison of two entirely separate matters. It is not enough to assume that these are different calculations of the same matter, because they are not. Historical volatility is derived from specific and readily observed closing stock prices and their statistical analysis. Implied volatility is based on estimates of where *future* volatility should be, given a set of assumptions that might or might not be accurate.

The widespread reliance on implied volatility (IV) in the options industry leads to the assumption that volatility leads price, when in fact it is the opposite. IV is nothing more than a sentiment indicator meant to demonstrate the market's perception of future volatility (but not direction of premium movement).

The calculation combines five segments of the Black-Scholes pricing model (see Chap. 10). These are the current option premium, the current stock price, the strike, time to expiration, and the assumed risk-free interest

---

[1] The calculated standard deviation is based on an average of 252 trading days per year. This is the term used to arrive at the annualized percentage of volatility.

rate. By adjusting the risk-free interest rate, different volatility levels can be accomplished.

The risk-free interest rate is an interesting concept. This is an assumed theoretical rate that can be earned with no risk of loss. To the extent that it is used in option pricing models, it is the assumed rate of return an investor could earn elsewhere if investing in different instruments. This usually translates to reliance on U.S. Treasury bond rates as "risk-free," even though the credit rating of U.S. government obligations was downgraded in 2011 and various agencies also downgraded Treasury debt in the years following.[6]

The downgrade of credit for U.S. debt changes a definition of risk-free. In the past, the "full faith and credit of the United States government" was the best guarantee available and conformed to a generally accepted definition of risk-free. However, since this entire discussion is based on a theoretical model, the downgrade has to be taken into consideration in determining whether or not U.S. Treasury securities truly are risk-free.

A practical definition of "risk-free" is elusive. Some economists have observed that risk is impossible to forecast, and thus, a risk-free rate cannot be directly observed or quantified.[7] In other words, any formula relying on a risk-free interest rate is based on guesswork and estimation, and not on specific or known interest rates.

A comparison between a calculated implied volatility and historical volatility is problematical. Large differences in the two calculations are meaningless as one (IV) is based on estimates and the other (HV) is based on known quantities in stock prices. If the purpose is to verify IV by analyzing a comparative outcome for two dissimilar calculations, why perform IV at all? With the inherent certainty of historical volatility, the bigger question should be whether it serves as a reliable indicator of market risk for options trading.

Implied volatility does not rely only on the sole variable of risk-free interest. It also relies on the variability of the underlying stock and the price of the option. As these are fixed values at the moment of the calculation, assumptions of future movement add exponential doubt to the accuracy of IV for determining the likely trend in an option's price.

Implied volatility also relies on a calculated premium value of options, the result of the bid/ask spread (difference between premium paid by buyers and credit received by sellers). The average of these two, the mid-price is commonly used in option pricing models such as Black-Scholes. Clearly, however, the fair price of an option depends on whether a trader is long (buying) or short (selling). The mid-price is merely an average of the two, and its use is inaccurate because buyers and sellers look at different sides of the pricing ledger. The larger the bid/ask spread, the greater the distortion in the pricing model. One study noted the misleading application of mid-price values in pricing models:

> Existing literature typically uses the quoted bid-ask midpoint as the option premium, but I show that small price movements in very low-priced options can lead to large percentage increases in the bid-ask midpoint, while these price movements are still in fact less than the bid-ask spread itself. Therefore, in many cases, using the bid-ask midpoint as the option premium leads to a large positive return, while using the original ask and the subsequent bid leads to a negative return. One can debate the correct methodology, since trades are often struck between the bid and ask quotes. However, I argue for including the bid-ask spread for a realistic picture and note the dramatic effect this has on options returns.[8]

The most justified use of IV is that it measures market sentiment about option pricing and determines whether volatility is likely to rise or fall (based on the risk-free interest rate and other assumptions). The estimates further allow for calculation of probability that strike prices will be reached by stock price by expiration. Option traders may take comfort in being able to determine levels of probability in outcomes. However, since IV is based on perceptions and estimates, the calculation itself is questionable.

Fundamental Volatility Correlated to Stock Price Behavior

The flaws in implied volatility are easily revealed, especially in comparison to the readily quantified benefits of historical volatility. Beyond that comparison, the correlation between a stock price's historical volatility, and fundamental volatility of the organization, further supports the use of a two-pronged methodology: reliance on fundamental volatility and analysis to select stocks appropriate for options trading, and the use of historical volatility to time entry and exit.

The term "fundamental volatility" describes either macroeconomic factors or a company's financial trends; it also is used to describe credit risk and return on investment in assets. However, regarding options trading, fundamental volatility most accurately is related to the tendency of reported fundamental results over time to be more or less predictable. In an organization whose revenues and earnings are consistent over a decade, fundamental

Table 2.2 Fundamental outcomes, 10 years—prepared by the author

Year	Revenue ($ mil)		Earnings ($ mil)		Debt cap ratio	
	WMT	JCP	WMT	JCP	WMT	JCP
2016	482,130	12,625	14,694	−513	30.7	76.9
2015	485,651	12,257	16,078	−771	31.5	73.3
2014	476,294	11,859	15,878	−1388	32.5	60.9
2013	469,162	12,985	16,999	−985	30.5	47.5
2012	446,950	17,260	15,766	−152	34.7	40.4
2011	421,849	17,759	15,355	378	33.9	36.2
2010	408,214	17,556	14,414	249	30.1	36.7
2009	405,607	18,486	13,254	567	30.0	45.8
2008	378,799	19,860	12,884	1105	29.1	34.1
2007	348,650	19,903	12,178	1134	32.5	41.2

Source S&P Stock Reports

volatility is low. In another organization with erratic increases and decreases in these outcomes each year, fundamental volatility is high.

The levels of fundamental volatility (in the sense of financial trends reported by the company on its income statement and balance sheet) can be observed by comparison. Investors naturally tend to seek out companies whose fundamental results are predictable and steady over time. Using three tests of volatility (revenues, earnings, and debt capitalization ratio), the relative level of year-to-year fundamental volatility is observable. For example, comparing Wal-Mart (WMT) to J.C. Penney (JCP), annual fundamental volatility in these three results is revealing, as summarized in Table 2.2.

On this table, the differences in fundamental volatility are glaring. To express the degree of change in outcomes from year to year, subtract each year's total from the previous year; and then calculate the percentage of change. The formula:

$$(C - P) \div P = \%$$

C   current year
P   past year
%   percent of change

For example, Wal-Mart's 2016 revenue of $482,130 (in millions) and the 2015 result of $485,651 are used to calculate the percentage of change with this formula:

$$(\$482,130 - \$485,651) \div \$485,651 = -7.3\%$$

Applying this formula for each year, the annual percentage of increase or decrease can then be expressed on a table and compared between companies. Table 2.3 compares revenue, earnings and debt capitalization ratio changes between Wal-Mart and J.C. Penney.

The differences, based on these trends, point out that as a measurement of risk, the mathematical calculation of annual percentage changes of key fundamental indicators adds to the understanding of how fundamental volatility directly affects a stock's historical volatility. In viewing the percentage of annual changes from year to year, to the stock price history, the correlation is evident, although not always direct. The Wal-Mart 10-year chart in Fig. 2.2 traces the prices from one year to the next, with revenue percentage changes indicated for each year. The overall trend reveals a growing price per share over the decade, accompanied by single-digit changes (all but the last year on the positive side) for the same period.

In comparison, J.C. Penney experienced much greater volatility. The price chart for 10 years is overlaid with changes each year in revenues, as shown in Fig. 2.3.

In the comparison between Wal-Mart and J.C. Penney, the differences are observable. Whereas WMT experienced positive revenue growth over a decade, JCP was on the decline. While the correlation is not exact, the overall relationship between fundamental volatility and stock price behavior appears on each chart. This outcome supports the argument that historical volatility and fundamental volatility are aligned more so than any connection established via the estimates of option prices based on implied volatility. As one in-depth study concluded, implied volatility tends to lack predictability, notably when it deviates excessively from the more precise outcome of historical volatility. In both forms of analysis, volatility tends to quickly revert to the mean, so expanded levels are likely to lead to distorted estimates in implied volatility. The reflection between fundamental history and historical volatility is a reliable method for stock selection among options traders, and also as a test of risk levels in the stock (which also translates to risk levels in the associated options).[9]

Fundamental volatility can be tested and compared to stock price trends in many different ways. The previous example was based on revenue trends over a decade. Another method involves the analysis of dividend trends (see Chap. 4). Those companies whose dividend is raised every year for at least 10 years (so-called "dividend achievers") tend to also report growing stock price levels over the same period—assuming that other fundamentals also support this level of growth. For example, as long as the debt capitalization ratio remains steady or declines, the increased dividend per share clearly is

Table 2.3 Change comparisons, WMT and JCP—prepared by the author

Year	Revenue change	
	WMT (%)	JCP (%)
2016	−7.3	3.0
2015	2.0	3.4
2014	1.5	−8.7
2013	5.0	−24.8
2012	6.0	−2.3
2011	3.3	1.2
2010	0.6	−5.1
2009	7.1	−6.9
2008	8.6	−0.2

Year	Earnings change	
	WMT (%)	JCP (%)
2016	−8.6	33.5
2015	1.3	44.6
2014	−6.6	−40.9
2013	7.8	−548.0
2012	2.7	−140.2
2011	6.5	37.0
2010	8.8	−56.1
2009	2.9	−48.7
2008	5.8	−2.6

Year	Debt capitalization ratio change	
	WMT (%)	JCP (%)
2016	−2.5	4.9
2015	−3.0	20.4
2014	6.6	34.9
2013	−12.1	17.6
2012	2.4	11.6
2011	12.6	−1.4
2010	0.3	−19.9
2009	3.1	34.3
2008	−10.5	−17.2

a positive fundamental trend. However, if the dividend increase is accompanied by an increased in the debt capitalization ratio (long-term debt as a percentage of total capitalization), the overall picture is extremely negative.

Price and revenue history, Wal-Mart

Fig. 2.2 Price and revenue history, Wal-Mart—chart courtesy of StockCharts.com

Price and revenue history, J.C. Penney

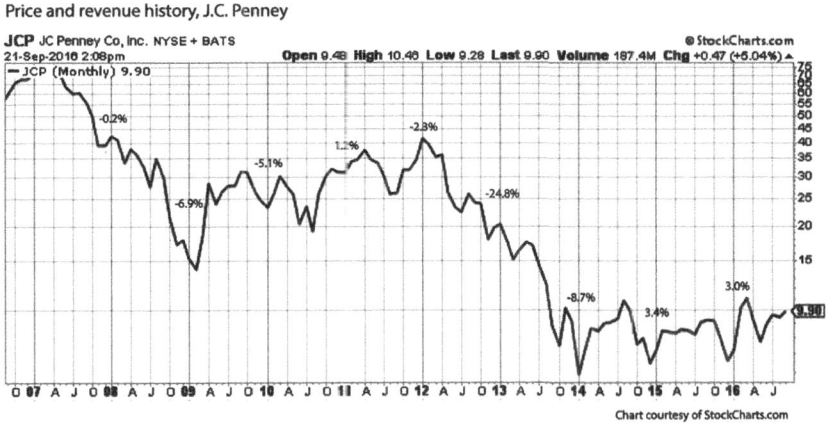

Fig. 2.3 Price and revenue history, J.C. Penney—chart courtesy of StockCharts.com

Bolstering dividends by acquiring higher long-term debt is a negative fundamental indicator.

The value of fundamental signals of many types, especially when correlated with historical volatility, provides stronger predictive intelligence concerning option risks than the less reliable estimates inherent in implied volatility. Another study based on analysis of option straddles confirmed this correlation, concluding that because

… fundamental signals contain information about future straddle returns that is incremental to what is captured in historical volatility, we expect higher hedge returns by combining historical volatility with fundamental signals …[10]

The Effect of Fundamental Volatility on Options Risk

Beyond the strongly observed correlation between fundamental volatility of a company and historical volatility of its stock price, it follows that a secondary question should be asked: Does fundamental volatility transfer into similar degrees of options risk?

This is an essential question given the widespread reliability on implied volatility to define options risk. Because IV is a flawed estimate of *future* risk levels, it does not provide any reliable measurements of actual options risk, only a flawed projection. So this leads to the question of how options risk should be defined. A strong case argues that historical volatility, especially when analyzed within a probability matrix such as Bollinger Bands, is a strong volatility measurement. (See Chap. 1.) Beyond that, the fundamental volatility of the company has a direct influence on option risk, just as it has been shown to directly influence historical volatility. Bollinger is based on the spread of two standard deviations, both above and below the middle band, so this version of historical volatility is broader than the alternative, normally based on a single standard deviation.

As a starting point, many studies have noted the effect of options on underlying stock prices. As options activity has been shown to influence a stock's price, the relative safety (volatility level) of options trades are correlated with not only the stock price but also with the fundamental volatility in the company. This interaction is unavoidable given the strong association between fundamental and historical volatility.[11]

Another analysis of this question noted the clear association between fundamental volatility and market risk:

> Our study proposes that fundamental volatility may be the correct measure of risk for the total market. Changes in fundamental volatility rather than observed volatility may be more appropriate for market regulators when they investigate the systematic effect of the introduction of derivatives on the market or the current state of the market. Regulators who currently compute the risk-neutral density of returns implied by option prices may wish to consider our procedure as a complimentary calculation to assess changes in the riskiness of markets.[12]

This observation concerning the nature of market risk is profound. To many options traders, the choice between fundamental analysis and technical analysis is a binary decision. A majority rejects fundamental analysis as dated and of no use in determining options risk. However, once it becomes apparent that fundamental volatility is correlated directly with historical volatility and, by association, with option pricing and risk, the value of fundamental analysis—even within the options market—is unmistakable.

The connection between fundamental trends and stock price behavior (historical volatility) has been observed through longer time periods as well. Plotting standard deviation of New York Stock Exchange-listed stocks revealed the highest levels of volatility were between 1929–1939 and during October, 1987. At these times, the same study concluded that stock market historical volatility was high in relation to fundamental values of companies.[13]

The reliance of historical volatility and its association with fundamental trends is clearly superior to any attempt at forecasting future volatility for options. This becomes important because over many years, attempts to develop accurate methods have failed:

> Despite their sophisticated composition, the predictive power of most volatility forecasting models is continually failing to convince investors of their designer's claims. Thousands of academics have devoted their entire careers to publishing models that supposedly are able to forecast volatility. Some authors have published well over 40 papers on this very topic ... and yet none seems to deliver any improvement over the simple standard deviation.[14]

The Proximity Factor

The use of historical volatility, calculated with the use of standard deviation, helps options traders to skillfully time trade entry and exit. However, another aspect to this requires yet another observation. The *proximity* of the current stock price to resistance or support increases the likelihood of reversal. There are five elements involved in this:

– historical volatility, with high levels favorable to reversal
– duration and angle of the trend, with stronger trends leading to stronger reversals
– strength of reversal and confirmation
– multiple confirmation
– price gapping to take price above resistance or below support

To reduce these five elements to a single statement:

> Reversal of price is most likely when historical volatility is high, when the duration and angle of the current trend is strong (fast price movement, sharp angle), when reversal signals are exceptionally strong and confirmed with equally strong signals (multiple signals is desirable), and especially when price gaps through resistance or support.

This description of ideal proximity encompasses all of the required elements: volatility, price, trend, reversal and confirmation. It is difficult to quantify, however, because the phenomenon varies with each stock and with its chart scale. A patient options trader recognizes the opportunity to exploit trends when all of these elements are present. Reversal should be timed in one of two ways. First, if the same elements as above appear indicating reversal in the opposite direction, the original trade should be exited and a new trend entered (replacing a bullish with a bearish trade, or a bearish with a bullish). Second, if a predetermined profit goal is reached. For example, if you are able to double the net value of the initial trade or accomplish a pre-set dollar amount of profit, a closing trade should be entered. After that, seek new proximity factors to enter a new trade.

Even though this set of elements is difficult to quantify, the set of requirements can be set up with a simplified mathematical evaluation in order to establish relative proximity values between two or more stocks. Table 2.4 provides guidance for this type of system.

While the selection of a rating for each of these elements is subjective, application to two or more situations overcomes the problem of dissimilar attributes on various charts and price patterns. For example, applying this test to two retail companies, Wal-Mart and J.C. Penney, reveals differences in the quality of proximity. The comparison is validated by applying the same standards to both stock charts.

The chart for Wal-Mart is shown in Fig. 2.4.

On this chart, a strong price move occurred in the third week of May. The price dropped well below the established trading range immediately before earnings were announced, in spite of two strong bullish candlestick signals. The overall signal value for this strongly pointed to the likelihood of a bullish reversal.

Applying the proximity ratings test, the results are summarized in Table 2.5.

2 The Role of Fundamental and Technical Analysis

Table 2.4 Rating system for proximity trade timing—prepared by the author

Description	Rating
Historical volatility:	
Highest, past 6 months	4
Highest, past 3 months	3
Trending higher	2
Not trending higher, or low	1
Duration and angle of trend:	
Rapid trend with sharp angle	4
Moderate momentum with medium angle	3
Slow momentum with low angle	2
Very slow momentum with very low angle	1
Strength of reversal and confirmation:	
Exceptionally strong signals	4
Moderate signals	2
Reversal without confirmation	0
Contradictory signals	−1
Multiple confirmation:	
3 or more confirmation signals were found	4
2 confirmation signals were found	2
1 confirmation signal was found	0
No confirmation signals were found	−1
Price gapping outside of trading range:	
Strong gapping move	5
Reversal occurs at resistance or support	3
Reversal occurs on approach to borders	1
Reversal occurs at the trend's mid-range	−1

Historical volatility adjusted to two standard deviations, expressed by way of the Bollinger Band width, was at three points. This was not high volatility, but was trending higher at that moment, thus justifying a 2-point rating.

The duration and angle of the trend was also given a 2-point rating based on the very slow move of the trend and its low angle.

Reversal and confirmation was exceptionally strong with a combination of two candlestick bullish reversals in close proximity to one another. These set up strong confirmation and result in the 4-point rating. The multiple confirmation added another 4 points.

Finally, the strong gap below the trading range set up an equally strong reversal, justifying the 5-point rating in the last category.

Wal-Mart stock chart

Fig. 2.4 Wal-Mart stock chart—chart courtesy of StockCharts.com

Added together, the 17 points out of a possible maximum of 21 represents an 81% reversal confidence:

$$17 \div 21 = 81\%$$

The system works well. Had this been applied at the moment the reversal occurred (on May 18 when the single gap below trading range and before formation of the morning star), the resulting bullish reversal was more likely to be anticipated and acted upon. The expanded chart, showing the next price move, is shown in Fig. 2.5.

In this example, the overall proximity strength combining all of the signals was calculated out to 81% confidence level and, as the subsequent price movement revealed, the results occurred as expected.

The same process could have been applied to the chart of J.C. Penney, shown in Fig. 2.6.

The point of interest on this chart is close to the end of the chart. Price gapped above resistance. However, the rising wedge is a weak bearish reversal,

Table 2.5 Rating system, Wal-Mart—prepared by the author

Description	Rating
Historical volatility:	
Highest, past 6 months	
Highest, past 3 months	
Trending higher	2
Not trending higher, or low	
Duration and angle of trend:	
Rapid trend with sharp angle	
Moderate momentum with medium angle	
Slow momentum with low angle	2
Very slow momentum with very low angle	
Strength of reversal and confirmation:	
Exceptionally strong signals	4
Moderate signals	
Reversal without confirmation	
Contradictory signals	
Multiple confirmation:	
3 or more confirmation signals were found	4
2 confirmation signals were found	
1 confirmation signal was found	
No confirmation signals were found	
Price gapping outside of trading range:	
Strong gapping move	5
Reversal occurs at resistance or support	
Reversal occurs on approach to borders	
Reversal occurs at the trend's mid-range	

and momentum is less than one point in the overbought region; so reversal signals are present but not strongly.

The rating for this situation is summarized in Table 2.6.

The stock trended higher, and was given a 3 rating based on the 3-month price history. Momentum is considered moderate and, even with the rapid jump above resistance, the larger bullish trend was moderate, thus the rating of 3 for the trend. The reversal lacked strong confirmation, so the strength was judged to be a 2-point rating. The confirmation signal was minimal, so multiple, confirmation was discounted and only 1 point was assigned. Finally, gapping action was strong so the final section was rated as a 5.

Overall, this adds up to 14 out of a possible 21 points:

Fig. 2.5 Wal-Mart stock chart, expanded—chart courtesy of StockCharts.com

Fig. 2.6 J.C. Penney stock chart—chart courtesy of StockCharts.com

Table 2.6 Rating system, J.C. Penney—prepared by the author

Description	Rating
Historical volatility:	
Highest, past 6 months	
Highest, past 3 months	3
Trending higher	
Not trending higher, or low	
Duration and angle of trend:	
Rapid trend with sharp angle	
Moderate momentum with medium angle	3
Slow momentum with low angle	
Very slow momentum with very low angle	
Strength of reversal and confirmation:	
Exceptionally strong signals	
Moderate signals	
Reversal without confirmation	2
Contradictory signals	
Multiple confirmation:	
3 or more confirmation signals were found	
2 confirmation signals were found	
1 confirmation signal was found	1
No confirmation signals were found	
Price gapping outside of trading range:	
Strong gapping move	5
Reversal occurs at resistance or support	
Reversal occurs on approach to borders	
Reversal occurs at the trend's mid-range	

$$14 \div 21 = 67\%$$

The outcome of 67% confidence was far lower than the Wal-Mart case, at 81%.

Based on these results, Wal-Mart's price advanced as anticipated by the bullish signals as well as the ratings system. In comparison, JCP moved up during August to $11.25 but declined by late September below $10 per share.

The proximity factor, expressed through the 5-part ratings system, works to a degree in anticipating the likelihood (but not the certainty) of short-term price trends. For options trading, this quantification of the probability for accurate forecasting improves the likelihood of well-timed trades.

Analysis of the underlying chart with the known valuation of historical volatility provides a compelling case for the timing of trades, and definitely more so than with the use of implied volatility. The next challenge is to determine and compare the pricing of options on a reasonable basis. Chapter 3 explores this topic.

Chapter Summary:

– the direct relationship between fundamental and historical volatility is easily proven
– historical volatility can be calculated with a simplified Excel worksheet formula
– historical volatility is precise, whereas implied volatility attempts to estimate future values
– fundamental volatility and stock price behavior are correlated directly
– fundamental volatility is further correlated to options risk
– proximity and the use of a rating system defines the probability of trading success.

Notes

1 Goodman, Theodore; Monica Neamtiu; & X. Frank Zhang (2012). Fundamental analysis and option returns. *The Hong Kong University of Science and Technology*. Abstract.
2 Roll, R., E. Schwartz, & A. Subrahmanyam (2010, April). O/S: The relative trading activity in options and stock. *Journal of Financial Economics*, 96, 1–17.
3 Rogers, J.; A. Van Buskirk & D. Skinner (2009, August 17). Earnings guidance and market uncertainty. *Journal of Accounting and Economics* 48, 90–109.
4 Guay, W. & J. Harford (1998, December). The cash-flow permanence and information content of dividend increases versus repurchases," *Journal of Financial Economics* 57, 385–415.
5 Daly, Kevin (2011, October). An overview of the determinants of financial volatility: An explanation of measuring techniques. *Modern Applied Science*, Vol. 5, No. 5.
6. Story, Louise (2011, August 17). U.S. inquiry is said to focus on S&P ratings. *New York Times*.
7. Tobin, James & Stephen A. Golub (1997). *Money, Credit and Capital*. New York: McGraw-Hill, p. 17.

8. McKeon, R. (2013). Returns from trading call options. *Journal of Investing*, 22(2), 64–77, 5.
9. Goyal, A. & A. Saretto (2013, April). Cross-section of option returns and volatility. *Journal of Financial Economics* 108(1), 231–249.
10. Goodman et al. *Op. Cit.*, p. 23
11. Conrad, J. (1989, June). The price effect of option introduction. *Journal of Finance* 44, 487–498.
12. Hwang, S. & Satchell, S. E. (2000). Market risk and the concept of fundamental volatility: Measuring volatility across asset and derivative markets and testing for the impact of derivatives markets on financial markets. *Journal of Banking and Finance*, 24, 759–785. p. 22.
13. Shiller, Robert J. (1989). *Market Volatility*. Cambridge MA: The MIT Press, pp. 95–96.
14. Daly, *Op. Cit.*

3
Pricing of the Option

Chapter Objectives:

- master put/call parity and its theoretical and practical applications
- identify parity's application in synthetic and straddle trades
- study the utility of upper and lower bound calculations
- analyze the differences in intrinsic, time and extrinsic value of options
- track methods to calculate Delta and Gamma and note how they apply to risk
- calculate option yields on several different stocks to spot the many variables involved.

Options pricing appears straightforward at first glance. However, when pricing levels are studied in detail, it becomes evident that an array of calculations are essential to arrive at an accurate conclusion about pricing.

Some believe in the Black-Scholes pricing model (see Chap. 10) to determine the fair price of an option. However, this is of questionable value for several reasons, including the multiple flawed assumptions that are involved with this model. On a practical level, option pricing is related directly to the options listing and the ever-changing prices; parity; bid/ask spread; how these affect timing of trades; and identifying high and low volatility.

The issue surrounding how to price an option often rests on a misleading assumption. Many options traders believe that the value of an option relies on expected value and changes in value of the underlying stock. If the stock price rises, a long call gains in value and a short put loses. If the underlying price falls, a long put gains in value and a short call loses.

Although underlying price movement obviously does affect observed changes in intrinsic value of the option, that is an outcome. For the more immediate question of how options are priced, "... the key characteristic of the underlying stock that concerns us is its volatility."[1]

This does not contradict the observation that implied volatility of the option is less reliable than historical volatility of the underlying. The underlying stock's historical volatility is the factor directly creating option volatility, and it is based on specific valuation. In comparison, implied volatility is the result of estimates. Thus, historical volatility is a starting point for how options are priced.

Affecting this is the level of interest among traders in a particular option itself. This is directly impacted by the moneyness of the option as well as by time remaining until expiration. These variables are expressed in put/call parity as well as by limitations seen in upper and lower bounds of option values. Put/call parity is defined in calculations of forward contracts in a particular way that brings in present value and assumed risk-free interest. For the purpose of judging the viability of entering a particular type of trade, put/call parity is defined in a raw and more basic observation of the pre-trading cost premium of options and their bid/ask spread.

Put/Call Parity

In options trading, put/call parity has two definitions. First, it can be used to describe the similarity in the bid/ask spread and the net debit or credit set up for any options trade employing different options with the same strike. This may refer to a combination of calls and puts, long and short, or both. The concept of put/call parity was introduced by Hans Stoll in 1969. In his paper, Stoll observed that "... relative put and call prices are closely related regardless of stock price movements or expectations about future movements."[2]

Varying premium values between calls and puts are commonplace. This is true whether the strike as at the money or close to it. Whenever the premium values are identical, they are at parity. The value of put/call parity applies in the use of specific strategies, notably for synthetic long or short stock or straddles. In these strategies, the existence of parity at the money is desirable.

For example, the synthetic stock strategy includes the purchase of a call and the sale of a put at the same strike (synthetic long stock), or the purchase of a put and the same of a call (synthetic short stock). A synthetic is so called because the combined option positions mirror price movement in the underlying point for point. At parity, the only cost should be trading fees, due to parity between the long and short option. The same observation applies to straddles, in which a long call and long put are opened at the same strike (long straddle); or a short call and short put are opened at the same price.

Because bid and ask prices often vary considerably between different stocks, parity is not always available. The use of strategies with different options at the same at-the-money strike may involve a small credit or debit adjustment. Parity often applies to forward contracts rather than to options; however, in options trading, parity brings attention to the relationship between bid and ask prices and the significance of the spread between the two. As a general rule, traders can readily identify low volume in a particular underlying's options by an exceptionally wide bid/ask spread. When the spread narrows as volume grows, it is a symptom of increased interest among traders in the particular option series. Whether this interest is on the long side or the short side is not readily identified until the moneyness of each option is examined, as well as the time remaining to expiration.

Due to the differences between the ask price (price to be paid for purchasing an option) and the bid price (price to be received upon sale of an option), the bid/ask spread should be analyzed on the basis of these two prices, especially in the case of synthetics, where one long and one short option are included. For straddles, the same distinction applies (ask for long straddles, and bid for short straddles). Using the example of Microsoft (MSFT) as of the trading session on September 29, 2016, at a point where the stock price was $58.01, an analysis of options with a 58 strike reveals how put/call parity is summarized.

Example for a synthetic long stock trade.
A trader expecting the stock price to rise may employ a synthetic long stock trade based on the 58 strike. The next question is whether to use a soon-expiring expiration or a longer-expiring expiration. Three examples:

October 14 (15 days):
    Buy 58 call, ask 0.69, plus trading fees, total cost $78
    Sell 58 put, bid 0.63, less trading fees, net cost $54
        Net debit $24

October 21 (22 days):
    Buy 58 call, ask 1.50, plus trading fees, total cost $159
    Sell 58 put, bid 1.45, less trading fees, net cost $136
        Net debit $23

October 28 (29 days):
    Buy 58 call, ask 1.59, plus trading fees, total cost $168

> Sell 58 put, bid 1.53, less trading fees, net cost $144
> Net debit $24

These three examples demonstrate that parity did not exist between the call ask and the put bid. The net debit in each case represented the spread plus trading fees (estimated at $9 for single option contract trades).

Example for a synthetic short stock trade.

On a short trade, combining a long put and a short call, the bid/ask prices vary in a different way due to the reversal of long and short for each position. Three examples based on Microsoft at the money:

> October 14 (15 days):
> Buy 58 put, ask 0.66, plus trading fees, total cost $75
> Sell 58 call, bid 0.66, less trading fees, net cost $57
> Net debit $18
>
> October 21 (22 days):
> Buy 58 put, ask 1.46, plus trading fees, total cost $155
> Sell 58 call, bid 1.49, less trading fees, net cost $140
> Net debit $15
>
> October 28 (29 days):
> Buy 58 put, ask 1.56, plus trading fees, total cost $165
> Sell 58 call, bid 1.56, less trading fees, net cost $147
> Net debit $18

In this example, entered on anticipation of a downward trend, parity was discovered in the first and third examples between the put ask and call bid prices. As a result, the net debit was equal to the combined $9 trading fee for each option, or a total of $18. The middle expiration was reduced by 3 cents due to the 3-cent difference between put ask and call bid.

Example for a long straddle trade.

> October 14 (15 days):
> Buy 58 call, ask 0.69, plus trading fees, total cost $78
> Buy 58 put, ask 0.66, plus trading fees, total cost $75
> Total debit $153
>
> October 21 (22 days):
> Buy 58 call, ask 1.50, plus trading fees, total cost $159

Buy 58 put, ask 1.46, plus trading fees, total cost $155
Total debit $314

October 28 (29 days):
Buy 58 call, ask 1.59, plus trading fees, total cost $168
Buy 58 put, ask 1.56, plus trading fees, total cost $165
Total debit $333

In these cases, the combined long positions create a middle "loss zone" equal to the total cost of the options. The breakeven point is equal to the distance either above or below the strike. If price moves above the high breakeven or below the low breakeven, a profit is earned. For example, in the October 28 expirations, the total cost was $333, so there are two breakeven prices:

High breakeven: 55 + 3.33 = $58.33
Low breakeven: 55 − 3.33 = $51.67

Example for a short straddle trade.

October 14 (15 days):
Sell 58 call, bid 0.66, less trading fees, net credit $57
Sell 58 put, bid 0.63, less trading fees, net credit $54
Total credit $113

October 21 (22 days):
Sell 58 call, bid 1.49, less trading fees, net credit $140
Sell 58 put, bid 1.45, less trading fees, net credit $136
Total credit $276

October 28 (29 days):
Sell 58 call, bid 1.56, less trading fees, net credit $147
Sell 58 put, bid 1.53, less trading fees, net credit $144
Total credit $291

In these cases, the combined short positions create a middle "profit zone" equal to the net credit of the options. Above or below this zone, the position loses. The breakeven point is equal to the distance either above or below the strike. As long as the underlying price remains within this zone, the overall position will experience a limited profit; if price moves above the upper breakeven or below the lower breakeven, a net loss will be experienced. For example, in the October 28 expirations, the total credit was $291, so there are two breakeven prices:

High breakeven: 55 + 2.91 = $57.91
Low breakeven: 55 − 2.91 = $52.09

The role of bid/ask spread and put/call parity affect these outcomes. Calculation of profit and loss zones are essential in order to select strategies with reasonable chances for favorable outcome, given personal risk tolerance and the degree of distance between the strike and the profit target.

Put/call parity has a more expanded application and calculation when computed precisely and based on inclusion of present value. In this variation of put/call parity, the calculation is based on European expiration of both call and put. The purpose is far beyond the one described above, where the purpose is to determine whether parity makes a particular strategy more desirable or less so.

In calculating put/call parity under the interest rate assumption, an adjustment is included to account for the option price and time until expiration as well an assumed risk-free interest rate:

$$C + X \div (1 + r)^t = P + U$$

C	premium of the call
X	strike
R	assumed interest rate
t	time to expiration
P	premium of the put
U	underlying price

This more widely known definition of put/call parity is applied for arbitrage trading and similar strategies to generate riskless profits. It is explained accurately in the following:

> In a competitive financial market, we expect that a combination of assets that leads to an equal future cash flow will have an equal market value. If the value of one combination were higher than the other combination, arbitrage opportunities that create immediate profit would exist. By buying the cheaper combination, while selling the expensive combination, we create riskless profit.[3]

For a majority of options traders, the spread between call and put premium received or paid (distance from parity in the immediate moment), is adequate for the purpose of testing a strategy and determining whether or not the likelihood of profitable results justify entry. However, anyone interested in spotting arbitrage opportunities will be likely to apply this more advanced formulation.

The interest rate assigned to this calculation is not the only potential variable. Many calculations used in pricing options or to identify probability are based on the use of either the mid-price between bid and ask, or on the last traded price. This is troubling since both versions of "price" are inaccurate. Because buyers and seller trade on opposite ends of the bid/ask spectrum, a calculated premium value should be calculated separately for buyers and sellers and based on the latest known actual bid or ask, not an average of the two. This issue rarely is raised in the explanation of how the volatility calculations, probability, or pricing model evolves with bid and ask, even when the spread is significant. Although the average is always a mid-point, the outcome is distorted, and more distortion is found when the spread is exceptionally wide.

To add accuracy to the calculation of bid/ask parity, the premium used should be more accurately defined as *either* bid or ask, depending on whether the option is going to be bought or sold. This would improve the calculation considerably in line with what occurs in a trade. Complicating this further, the formula including time and interest works only for a European option, which cannot be exercised until close to expiration. It works for an American style option only if the contract is kept open until expiration, which occurs only in a minority of cases. Finally, the introduction of dividends paid on stock also requires adjustment to a pricing formula.

In conclusion, the many variables involved with attempting to calculate put/call parity based on time and interest as well as premium values is easily distorted by using the wrong premium (bid versus ask); and the entire matter becomes purely academic if the option is closed before expiration. These variables support the more basic and easily observed version of put/call parity based on the examples for synthetics and straddles.

Upper and Lower Bounds

The *theory* concerning maximum extent of option price movement is referred to as upper and lower bounds. This is theoretical because it represents a mathematical attempt to define a likely range of value for a particular option.

An option can never be worth more than the underlying stock, so the maximum upper bound is equal to the underlying price. Anyone who has traded options immediately recognizes the low likelihood of an option achieving that level of value. However, as a finite definition of the upper bound, it places a limit on potential value. The formula for this is expressed as:

$$C \leq S$$

C value of a call
S current price of stock

The call's premium must be equal to or less than the price of stock, identifying the maximum level, or upper bound of the option's price. On the other side of this equation is the lower bound of a put. The equation is different, as the put may never be worth more than the strike price. Thus, the put's lower bound formula is:

$$P \leq S$$

P value of a put
S strike price of the option

These definitions have to be adjusted for two additional distinctions. First is the attribute of exercise. An American option can be exercised at any time before expiration, whereas a European option can be exercised only at or near the expiration date. Second is the inclusion or exclusion of a dividend, which affects the range of price within upper and lower bands.

The theory behind definitions of upper and lower bands is useful information. However, for the day trader or swing who tends to move rapidly in and out of option positions, the application of these bounds is not immediately important in the timing of a trade. An options trader knows rationally that the bounds exist and that the range of possible prices conform to those bounds. Expressed another way:

Price bounds for a call $= Z, U - E$
Price bounds for a put $= Z, E - U$

Z zero
U underlying price
E exercise price (strike)

Intrinsic Value

For the purpose of bringing order to the definition of *premium* for an option, the overall value has to be divided into its individual parts. Most descriptions of this issue define premium as having two parts, intrinsic value and time value. However, an accurate definition of time value relates strictly to time itself, and not to volatility. In using only these two distinctions, time value rises or falls as implied volatility changes. An accurate model recognizes a third segment of value, which consists of implied volatility (IV), also termed extrinsic value.

Although the reliance on implied volatility is questionable as a means for timing of options trades (with preference for the more precise historical volatility), the existence of IV affects the overall premium value of the option. Often overlooked, however, is the fact that IV is likely to track historical volatility very closely. Thus, it is usually accurate to describe extrinsic value as the volatility effect, whether implied or historical.

By separating premium into three segments, overall pricing is more readily comprehended and more accurately tracked. The complete picture of an option's premium consists of intrinsic, time and extrinsic value.

Intrinsic value describes the degree to which an option is in the money (ITM), meaning the underlying price is higher than a call's strike or lower than a put's strike. The formula for ITM options depends on the type of option:

$$U - S = C_I$$

$$S - U = P_I$$

U underlying price
S option's strike
C_I call ITM
P_I put ITM

The formula assumes a positive result in order for intrinsic value to exist. If the strike is equal to the underlying price, the option is at the money and there is no intrinsic value. When the underlying is lower than the call's strike or higher than the put's strike, the option's condition is out of the money and there is no intrinsic value.

The formula for out of the money (OTM) condition is the opposite. The lack of intrinsic value defines OTM condition for an option:

$$S - U = C_O$$

$$U - S = P_O$$

U underlying price
S option's strike

C_O call OTM
P_O put OTM

The formula for at the money identifies the same level for strike and underlying:

$$S = U = C_A$$

$$S = U = P_A$$

U underlying price
S option's strike
C_A call ATM
P_A put ATM

The moneyness of an option can be visually summarized as shown in Fig. 3.1.

Time Value

In spite of the popular combination of time and volatility to describe option pricing as having two parts, a more accurate and descriptive method is to separate *time* and *volatility* and define these as separate types of premium.

Time value is just as predictable as intrinsic value. However, the specific dollar amount of intrinsic value is easily identified. It is equal to the number of points the option is in the money. A two-point distance between strike and underlying price established $200 in intrinsic value. When the underlying price is $47 and a call's strike is 45, the call contains $200 of intrinsic value; likewise, when the underlying price is $47 and a put's strike is 50, the put contains $300 of intrinsic value.

Time value is related solely to the amount of time remaining until expiration. With many months remaining in an option's life, time value decreases slowly. As expiration approaches, time decay accelerates. In the final month before expiration, and especially in the final two weeks, time decay as at its most rapid. This curve of acceleration in time decay is shown in Fig. 3.2.

When time and extrinsic value are combined as a single form of premium, calculation of time value is quite complex. It also varies with the moneyness of the option. However, when time value is analyzed by itself, the time decay effect on option premium is observable as related to time alone.

Moneyness of an option

Fig. 3.1 Moneyness of an option—chart courtesy of StockCharts.com

The dollar value of theta can be estimated by first estimating the degree of volatility in premium (extrinsic value), then noting the remainder as time value:

$$P - I - V = T$$

- P total premium
- I intrinsic value
- V implied volatility
- T time value

Time value, when isolated apart from extrinsic (volatility) value, is accurately defined as the rate of an option's declining value over time. The dollar value varies based on overall premium value and this is a factor of the underlying price. For example, a $100 stock will contain a higher dollar level of time value than a $10 stock. The rate of decline in time value is predictable, however, and can be further separated from implied volatility by observing the remaining time to expiration. The effect of time decay is most easily observed in the last month of an option's life.

Some option strategies exploit rapidly declining time value by focusing on short positions close to expiration. In the final two weeks, for example, even as a short option moves in the money, it is possible to experience an offset to intrinsic value due to time decay. Even as the intrinsic dollar value increases, the overall premium may decrease, reflecting the natural acceleration of time decay even at a level greater than the increase in intrinsic value.

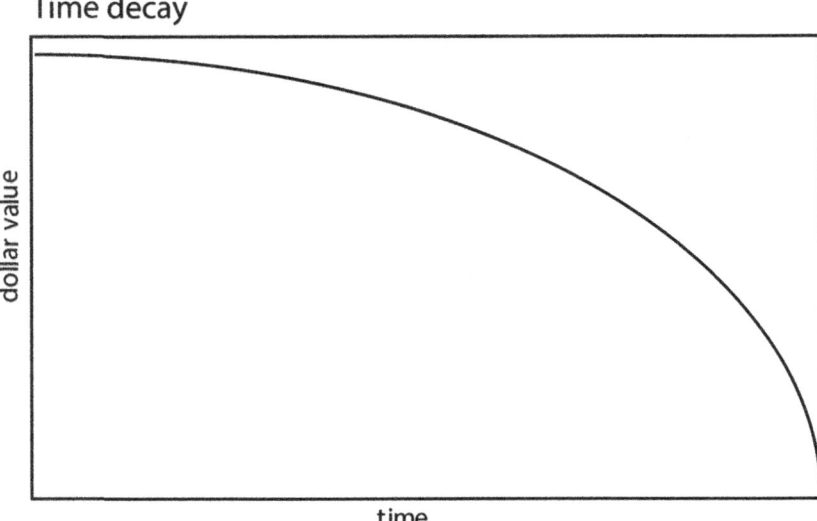

Fig. 3.2 Time decay—prepared by the author

Extrinsic Value (Volatility)

The study of implied volatility has to include estimates that attempt to identify future premium values. While volatility exists and can be articulated in today's option premium, it is often mischaracterized as a *predictive* force. Volatility, by definition, reflects recent price behavior and does not contain any predictive powers. A recurring criticism of IV has been that when it varies from levels of historical volatility, it often points to mispricing of the option. In comparison, realized volatility—the tendency of stock price ranges to increase as a reflection of perceived risk—is more dependent on the price trend seen in historical volatility:

> Implied volatility is widely believed to be informationally superior to historical volatility, because it is the "market's" forecast of future volatility. But for S&P 100 index options, the most actively traded contract in the United States, we find implied volatility to be a poor forecast of subsequent realized volatility. In aggregate and across subsamples separated by maturity and strike price, implied volatility has virtually no correlation with future volatility, and it does not incorporate the information contained in recent observed volatility.[4]

This problem points out that implied volatility, while part of overall option premium, is not of value in estimating future pricing of an option's premium. The value in implied volatility is in how it can be correlated with historical volatility, which is both specific and measurable. When IV levels are inconsistent with historical volatility, it indicates that the option is mispriced, at least in the moment. Such inaccuracies tend to revert to the mean rapidly, so recognizing mispricing level points to trading opportunities that are likely to evaporate rapidly.

Using IV to improve timing of options trades is a popular pursuit. An assumption that IV and historical volatility will be similar in most instances may lead to the use of visualized volatility levels, such as the demonstrated application of Bollinger Bands in comparison to underlying price levels and to option premium. As volatility increases (as measured by the two standard deviations in Bollinger Band width, for example), option premium is also expected to increase, most notably for soon-to-expire at-the-money contracts. In this application, the similar between IV and historical volatility causes changes in volatility premium. On a comparative basis, this is observable in increasing or decreasing overall premium that cannot be explained other than on the assumption that volatility is responsible. Thus, volatility is in effect the "risk premium" of the option.

When a change in overall premium occurs that is not correlated in the easily observed historical volatility of the underlying, it points to the likelihood of mispricing. Because these errors quickly revert, it also points to momentary trading opportunities. However, attempting to exploit assumed mispricing, either through short-term swing trading, hedged positions or arbitrage, relies on the assumption that volatility is predictable. In practice, option pricing might reflect other factors not immediately apparent in either stock or option prices. Assuming the efficient market theory is at work, in which all known information is immediately reflected in price, the possibility for a misleading short-term trend cannot be overlooked. Because the markets are "informationally efficient," implied volatility will move according to the same forces.

However, a great flaw in this informationally efficient market is that it treats all information in the same manner. Thus, accurate information is treated in the same efficient manner as rumor, gossip or unsubstantiated claims about a company. This "informationally efficient" market is not necessarily "economically efficient" at the same time. Another observation concerning efficiency is that its existence is questioned broadly by the market at large (and especially by options traders). A completely efficient market would discourage and even prevent and speculative trading:

The paradox of efficient markets is that if every investor believed a market was efficient, then the market would not be efficient because no one would analyze securities. In effect, efficient markets depend on market participants who believe the market is inefficient and trade securities in an attempt to outperform the market.[5]

Implied volatility is flawed because it relies on estimates, which contradicts a presumption of efficiency in the markets. As part of the well-known Black-Scholes pricing formula, with its numerous flawed assumptions (see Chap. 10), IV can be best understood by analysis of changing option premium levels based on changing levels of historical volatility (HV). As HV changes (especially with rapidly changing levels during short-term rapid price movement), a corresponding change in overall option premium will be observed immediately. Discounting the obvious and observable impact of intrinsic value as options move in the money during times of dynamic price trends, the remaining option premium should be expected to reflect higher volatility levels when trading breadth expands, and lower volatility levels when it contracts. This applies whether a trader relies on estimates in IV or precise patterns in HV.

Estimating Delta

Expanding on the overall premium analysis of options, and especially on volatility (where all uncertainty of pricing occurs), the option's *delta* is a descriptive attribute defining how option pricing reacts to movement in the underlying. The moneyness of the option further determines how option pricing is likely to react to historical volatility. This is logical, since moneyness is perhaps the most important attribute of cause and effect (between underlying volatility and option value).

Delta, the Greek letter Δ, is used to identify this relationship. A useful application of Delta is to identify the likely levels found at various moneyness levels of an option. The Delta value ranges between 1.0 and -1.0. Given the upper and lower bounds of an option, Delta goes a step further by identifying how premium value behaves and varies based on the distance from the current underlying price to the option strike. This rule-of-thumb approach identifies Delta levels as summarized in Table 3.1.

Although these are estimates of expectations for Delta, they are useful in understanding the interaction between historical volatility and option premium. It would not be realistic to expect all options to react

Table 3.1 Approximation of delta for long options—prepared by the author

Strike	Long call moneyness	Call delta	Long put moneyness	Put delta
15	Deep ITM	1.0	Deep OTM	0
20	Next deep	0.9	Next deep	−0.1
25	Nearest ITM	0.75	Nearest OTM	−0.25
30	ATM	0.5	ATM	−0.5
35	Nearest OTM	0.25	Nearest ITM	−0.75
40	Next deep	0.1	Next deep	−0.9
45	Deep OTM	0	Deep ITM	−1.0

Current underlying value = $30

to volatility in the same manner, since moneyness is crucial to valuation of a particular option. The formula for Delta consists of dividing the derivation (∂) of the option (O) by the derivation of the underlying stock (S):

$$\Delta = \partial O \div \partial S$$

The application of Delta in recognizing or anticipating option premium movement helps to better understand how various strikes are likely to change given a move in the underlying. An ATM option is likely to change by 0.5 points for each point of movement in the underlying. Thus, a two-point upward move in the underlying should cause a one-point upward move in the option's correspond ATM premium:

$$2 * 0.5 = 1.0$$

A put at the same ATM status would be expected to change by 0.5 points. Since the put's Delta is negative, a two-point upward move should cause a one-point decline in the put's value:

$$2 * -0.5 = -1.0$$

As Delta levels change, so will the corresponding reaction to price, noting that the farther away from the money, the more significant the change. The deeper in the money, the more reaction to expect, all the way up to 1-to-1 for deep ITM, and close to zero movement for deep OTM.

These general "rules" will vary due to other factors, notably the time to expiration. As time moves closer to expiration, Delta accelerates for ATM or near-ATM options. Another influencing factor is volatility itself. As volatility increases, Delta (especially ATM strikes) tends to change, either in a positive or negative way depending on perceptions about likely direction of that underlying change.

Just as the estimated Delta values are opposite for long calls and puts in the preceding example, they are also opposite for short positions and, in fact, are flipped. Table 3.2 summarizes the difference for short positions. The positive and negative are reversed for short options in comparison to those for long options in the previous table.

The study of Delta is useful in determining how many options would be needed to hedge a position in the underlying. When the two sides are at parity, it sets up *position Delta*, also referred to as a *hedge ratio*: "Delta is a measure of the dollar change in an option resulting from a dollar change in the value of the underlying asset. It is an extremely useful option-pricing statistic, being a prerequisite for the determination of an option hedge ratio."[6]

For example, an ATM option with Delta of 0.5 identifies a 50% chance of that option ending up positioned ITM, and equally, a 50% change of its ending up OTM. When viewed in this manner, Delta is a proxy for probability based on the moneyness of the option. Thus, two ATM options are required to hedge a corresponding position in the underlying. Two long options would hedge one short option with maximized Delta of 1.0:

$$(2 * 0.5) = 1.0$$

This Delta neutral relationship can be adjusted to position Delta that is either positive or negative, by altering additional long or short positions. A bullish adjustment involves adding more calls. Thus, moving from two long calls to three sets of a Delta positive relationship of 1.5 rather than the exact hedge of 1-to-1:

$$(3 * 0.5) = 1.5$$

The same hedging adjustment reduces a ratio on the put side to create a Delta negative for a bearish stance in the options (either with short calls or long puts). Because the range of Delta is between zero and one, the mathematical relationship between call and put Delta is expressed as:

$$\Delta_p = \Delta_c - 1$$

Δ delta
P put
C call

A comparison of call and put values on the previously introduced Table 3.1 demonstrates this relationship at every level.

Table 3.2 Approximation of delta for short options—prepared by the author

Strike	Short call moneyness	Call delta	Short put moneyness	Put delta
15	Deep ITM	−1.0	Deep OTM	0
20	Next deep	−0.9	Next deep	0.1
25	Nearest ITM	−0.75	Nearest OTM	0.25
30	ATM	−0.5	ATM	0.5
35	Nearest OTM	−0.25	Nearest ITM	0.75
40	Next deep	−0.1	Next deep	0.9
45	Deep OTM	0	Deep ITM	1.0

Current underlying value = $30

Without doubt, volatility plays a role in setting parameters for hedge ratios. For this reason, a second Greek calculation, Gamma, is required in order to test the rate of change in Delta.

Estimating Gamma

Another Greek calculation, Gamma (symbol Γ), measures the rate of change in Delta and is expressed as a percentage. Gamma is also referred to as the option's curvature because it measures the rate of change in Delta. However, like Delta, the Gamma value is an estimate and should be used only to approximate the degree of relationship between Delta and the underlying.

Because it measures the acceleration of change in an option's directional premium trend, it is also a proxy for risk. Just as historical volatility represents levels of risk, a high Gamma implies higher risk and a low Gamma represents low risk. However, it measures the option's Delta rather than directly tracking historical volatility of the underlying. One observation making this of interest in trading options is that when Gamma is high, changes in pricing tend to occur rapidly.

The inherent trading risk associated with price changes (volatility) define whether a particular trade is an appropriate match for a given risk tolerance. This cannot be known without understanding the role of gamma. This, the *volatility risk premium* becomes a more severe factor in outcomes as volatility itself increases:

> … ignoring an option's gamma can lead to incorrect inference on the magnitude of the volatility risk premium … the S&P options are used as a test case to demonstrate the impact of ignoring gamma on the estimation of the

market price of volatility risk. The findings show that the more prices fluctuate, the greater the variability in the estimation of volatility risk premium when gamma is ignored.[7]

Gamma tends to have the highest percentage value when the option is at or close to the money; and will be at its lowest when the option is deep in or out of the money. Gamma for ATM options also tends to increase as expiration draws closer. However, at the same time, deep ITM or OTM Gamma decreases.

In addition to moneyness and movement in Delta, time to expiration and volatility further affect Gamma and its changing levels. Low volatility translates to higher Gamma when the option is ATM, and to lower Gamma (approaching zero) for deep in or out of the money. High volatility, in comparison, tends to reflect similar Gamma levels at each strike. This is a reflection of how moneyness of the option responds to movement in Delta which, in turn, is affected by historical volatility. As expiration approaches, this relationship becomes crucial, notably when volatility is low (meaning less likelihood of dramatic premium changes as expiration approaches).

This relationship between historical volatility and Gamma is summarized in Fig. 3.3.

This chart demonstrates how Gamma behaves as volatility changes. This often is referred to as a reflection of implied volatility. However, since Delta changes based on the underlying (and Gamma is related to Delta), both of these indicate the condition of historical volatility, or actual changes in the underlying rather than estimates of future changes in the option.

Gamma also behaves according to time remaining until expiration. A short-term option tends to report exceptionally high Gamma, and in comparison, a 6-month option will experience little movement over time. As a result, the interaction between historical volatility and time decay is observable based on the Gamma trend as well as option premium. The premium level drops as time decay accelerates. Gamma combines Delta behavior with time to establish the timing for trades with historical volatility in mind.

When options are ATM, Delta is 0.5 for calls and −0.5 for puts. At this point, Gamma is at its highest level, especially for soon-to-expire contracts. It is expressed as a positive for long options or as a negative for short positions, and is identical for calls and puts. A basic formula for Gamma is calculated as the second derivation (∂^2) of an option's value, divided by the second derivation of underlying price:[8]

$$\Gamma = \partial^2 V \div \partial^2 S$$

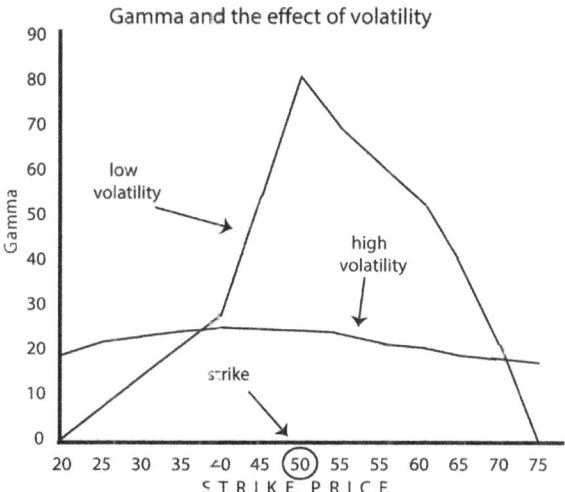

Fig. 3.3 Gamma and the effect of volatility—prepared by the author

Γ Gamma
$\partial^2 V$ second derivation of the option
$\partial^2 S$ second derivation of the underlying

The second derivation is applicable because Gamma represents the rate of change in a separate level of change (option price in relation to underlying price).

Calculating Relative Option Yield

A final attribute of pricing of options is the calculation of premium itself, as a percentage of the underlying price. A tendency among traders is to focus on the dollar value of an option's current premium without regard for other key attributes, including moneyness, time to expiration, and the most basic attribute of all, the yield.

Complicating this is the variation of moneyness and time to expiration. With different moneyness at different underlying price levels, like-kind comparisons often are difficult if not impossible. For example, if a particular option has a current ask price of 3 ($300). The current price of stock is $85 per share. The yield is 3.5%:

$$3 \div 85 = 3.5\%$$

This is far from the final result of a calculation, especially for comparisons between two or more options. Consider the variables:

1. Moneyness of the option.
2. Time remaining until expiration.
3. Spread between ask and bid prices (and how this affects the differences in option yield between long and short entry).
4. Whether or not a dividend is paid, as dividends affect the total return of options.
5. Comparisons to stocks at much higher or lower price per share and option values with equivalent moneyness strikes (for example, with the stock at $85 and option at the next highest strike of 86, the equivalent distance for a stock priced at $850 should be the 860 option).

These variables point out when like-kind comparisons of yield present a particular challenge. One popular strategy is to sell short strangles with strikes two standard deviations above and below the current price of stock. This sets up a comfortable cushion, but it appears to work most effectively for higher-priced stocks, due to all of these variables. A study of two different stocks makes this point.

The first stock is Priceline (PCLN). The stock chart as of the close of October 4, 2016, with Bollinger Bands overlaid on price, is shown in Fig. 3.4.

Priceline closed on October 4 with a bandwidth of 77.55 points, the distance between Bollinger's upper and lower bands. This represents a span of two standard deviations from the center band.

Applying this range to the current stock price of $1489.02, the same two standard deviation range can be calculated by first dividing the total bandwidth in half:

$$77.55 \div 2 = 38.78$$

This distance is next applied to current price to arrive at the same two standard deviation extent from top to bottom:

$$1489.02 + 38.78 = 1527.80$$

$$1489.02 - 38.78 = 1450.24$$

To execute a short strangle extending two standard deviations, a trader will search for strikes at these approximate levels. Based on the October 21 expirations (17 days from the close), the following bid prices were discovered:

Priceline stock chart

Fig. 3.4 Priceline stock chart—chart courtesy of StockCharts.com

$$1,530 \text{ call bid } 9.40 \ (0.6\% \text{ of strike})$$

$$1,450 \text{ put bid } 10.80 \ (0.7\% \text{ of strike})$$

The second example is Intuitive Surgical (ISRG), which closed on October 4 at $716.31, close to 50% of the price per share of Priceline. The chart for ISRG is shown in Fig. 3.5.

ISRG spanned a Bollinger bandwidth of 68.21 points on the target date. This level represents two standard deviations from top to bottom when compared to the center band. Applying this to the current stock price of $716.31, the same two standard deviations can be calculated. The first step is to divide the bandwidth by two:

$$68.21 \div 2 = 34.11$$

Applying this distance to current price, the target strikes are revealed:

$$716.31 + 34.11 = 750.42$$

$$716.31 - 34.11 = 682.20$$

A short strangle is set up based on two standard deviations by approximating strikes at these levels. Using the October 21 expirations, the following positions were identified:

$$750 \text{ call bid } 6.10 \ (0.8\% \text{ of strike})$$

$$680 \text{ put bid } 6.00 \ (0.9\% \text{ of strike})$$

Intuitive Surgical stock chart

Fig. 3.5 Intuitive Surgical price chart—chart courtesy of StockCharts.com

The yields are within one-tenth of one percent of one another for each of these test cases. PCLN yielded 0.6% on the call and 0.7% on the put, and ISRG yielded 0.8% on the call and 0.9% on the put. These all resulted by dividing the premium dollar amount by the option's strike.

These two examples are comparable in the sense of setting up a buffer zone equal to the two standard deviation ranges both above and below price. However, while ISRG is less than one-half the price per share of PCLN, the Bollinger range is very close, with only seven points difference between the two Bollinger sets.

This disparity can be explained. The buffer zone is based on the statistical distance of two standard deviations above and below price, regardless of the fact that price levels are so dissimilar. In terms of historical volatility, the two examples are comparable because of the differences in recent historical volatility trends. PCLN's historical volatility was unchanged throughout the month of September, but ISRG's historical volatility increased by 3.5 times from an average of 20 points at the beginning of September to 68 points by the target date. Aan attempt to make these two issues truly comparable has to also consider differences in levels of volatility, not only on the target date but also in recent history.

This exercise reveals two key points. First, initial yields (dollar value of options divided by strikes) are comparable only if the time to expiration is identical and only if the historical volatility is used as the basis for discovering applicable strikes. Second, this comparison has to take into account the different prices per share and likely volatility.

The examples based on strikes removed by two standard deviations in both directions reveals that option yield is based on numerous variations, including moneyness. A study of ATM options makes this point. Rather than using deep OTM options, a comparison to ATM options shows how much the yield can vary between different underlying stocks. PCLN strikes closest to the current price level were the 1490 strikes:

1490 call bid 25.10(1.7%)

1490 put bid 25.00(1.7%)

In comparison, the ISRG strikes had a much different outcome using the closest ATM strikes:

715 call bid 19.30(2.7%)

715 put bid 17.60(2.5%)

The lower-priced ISRG options with strikes ATM yielded a full percentage point higher than the comparable options for PCLN. Part of this disparity is explained by differences between the two strikes. For PCLN, the 1490 strike is 0.98 points above the closing price. For ISRG, the 715 strike was 1.31 points below. Although both examples were for the strikes closest to the money, these variations complicate the comparison.

Traders who make comparisons such as this may rely on the calculation of two standard deviations, regardless of underlying price levels; and then calculate the net yield of options based on strikes. The outcomes will never present perfect comparisons. However, they do present a reasonable level of comparison based on the true price range (based on Bollinger Bands and two standard deviations to set the distance to appropriate strikes), and are also reasonably comparable (in this example, within one-tenth to two-tenths of a percentage point).

Both of the sample stocks were non-dividend paying companies, which further adds to the argument that these are comparable. When dividends are added into the calculation of options valuation, the process is more complex. The next chapter explores the role of dividends in selection of options.

Chapter Summary

- put/call parity has strong practical applications in trade selection
- parity is especially crucial for finding strikes in synthetic and straddle trades

- upper and lower bands clarify likely ranges of option price levels
- the three distinct forms of option premium each contain their own attributes
- Delta and Gamma identify option volatility and risk
- option yields have to take into account the many variables between underlyings.

Notes

1. Chriss, Neil A. (1997). *Black-Scholes and Beyond.* New York: McGraw-Hill. p. 120.
2. Stoll, H. (1969). The relationship between put and call option prices. *The Journal of Finance, 24*(5), 801–824.
3. Nissim, Ben David & Tavor Tchahi (2011). An empirical test of 'put call parity.' *Applied Financial Economics* 21, 1661–1664.
4. Canina, Linda & Stephen Figlawski. (1993). The informational content of implied volatility. *The Review of Financial Studies* 6(3): 659–681.
5. Nayak, Keyur M. (2008, October 22). A study of random walk hypothesis of selected scrips. *ASBM Journal of Management* 1.1: 118–127.
6. Strong, Robert A. & Dickinson, Amy (1994, Jan/Feb). Forecasting better hedge ratios. *Financial Analysts Journal, 50*(1), 70.
7. Doran, James S. (2007). The influence of tracking error on volatility risk premium estimation. *The Journal of Risk, 9*(3), 1–36.
8. Chriss, *Op. Cit.,* pp. 311–312.

4

The Dividend Effect

Chapter Objectives:

- Analyze the role and complexity of dividends in selection of options
- Track dividend trends side-by-side with long-term debt trends
- Study methods for accurately calculating total return
- Compare various holding periods with timing of quarterly dividends
- Estimate the reliability of assumed future dividends versus future options trades
- Evaluate the role of dividend trends in actual outcomes of options trades.

Options entry and exit may be timed on either implied or historical volatility. A case was offered in preceding chapters for the more reliable application of historical volatility and recognition of the related trends through visual charting methods (for example, overlaying Bollinger Bands to set up a probability matrix for the timing of trade). Rather than relying on estimates of future price, this system is based entirely on the current status of price behavior.

A case also was made for the direct correlation between fundamental trends and stock price trends (as well as on option trade timing). However, options traders recognize that fundamental trends are not necessarily correlated directly with options valuation. One fundamental indicator is an exception to this general observation. Dividends are correlated directly to the selection and valuation of options, and should also be included in the calculation of an option trade's total return.

Dividends as Fundamental Indicators

Investors selecting stocks for their portfolios are likely to base their selection on dividend yield. This is a popular indicator representing income apart from capital gains on the stock. However, selecting a company and its stock based only on dividend yield or dividends per share is not always a reliable symptom of a strong fundamental investment. By association, a high dividend per share is also not automatically a smart selection for options trades.

Dividend yield is incorporated into options profits to arrive at total return. Analysis reveals an interesting lack of correlation between yields on high-priced and low-priced common stocks. This clearly has ramifications for comparative total return on options. Traders tend to believe that higher-yielding common stocks also yield better total returns on options. This belief is not supportable. In a follow-up to the well-known Black-Scholes paper of 1973, a 1974 study concluded.

Debt capitalization ratio, that:

> it is not possible to demonstrate, using the best available empirical methods, that the expected returns on high yield common stocks differ from the expected returns on low yield common stocks either before or after taxes. A taxable investor who concentrates his portfolio in low yield securities cannot tell from the data whether he is increasing or decreasing his expected after-tax return by so doing. A tax exempt investor who concentrates his portfolio in high yield securities cannot tell from the data whether he is increasing or decreasing his expected return.[1]

The expected return noted in this paper is in reference to what a trader estimates from a trade, based on assumed rates of returns. It is based on historical returns of similar trades; however, as an estimate, expected return is not a reliable method for approximating actual outcomes. It can be used to evaluate the likelihood or probability of a positive outcome as part of a risk assessment, but it is never more than an estimate.

The flaw in over-reliance on expected return is based not only on the fact that it is only an estimate, but that changing risk levels cannot be known in advance:

> The expected market return is a number frequently required for the solution of many investment and corporate finance problems, but by comparison with other financial variables, there has been little research on estimating this

expected return. Current practice for estimating the expected market return adds the historical average in realized excess market returns to the current observed interest rate. While this model explicitly reflects the dependence of the market return on the interest rate, it fails to account for the effect of changes in the level of market risk.[2]

Even with its limitations, however, expected return reflects a correlation between levels of risk and potential profitability. A distinction is made between the hedging aspect of options trading within expected return, versus the consequences of changes in volatility. One study addressed this stating that:

> … by examining option returns we can consider the economic payoff for taking very particular risks. For example, we can estimate the expected return of a position that is essentially immune to both small market movements and the risk of large market crashes, but is very sensitive to changes in volatility.[3]

In this context, comparative calculations of expected return may enable an options trader to appreciate options pricing behavior, not so much in response to underlying price movement but to ever-changing historical volatility.

This does not add tangible value to expected return by itself, but can add insight to option pricing behavior, notably regarding total return. However, no indicator, including dividends, should ever be relied upon in isolation. The larger context may point to a weakening fundamental trend which includes high dividends per share and impressive dividend yield. A strong option candidate should contain not only exceptional technical features (related to historical volatility of the underlying, low bid/ask spread, and attractive option premium). It should also be strong fundamentally, and this is where the analysis of dividends may reveal either underlying strength or weakness.

An analysis of dividends per share is a popular method for determining fundamental value. Dividends, of course, also are taken into account for calculation of options total return, so this is perhaps the most important fundamental trend for options traders. When the dividend is increased every year for 10 years or more, the company is referred to as a "dividend achiever." However, this is not the entire story.

To analyze dividend growth accurately, the trend has to be considered in line with the trend in long-term debt. Competition among companies may include market pressure to increase dividends every year without fail, even if earnings growth does not justify an increase. This behavior can have a detrimental effect on financial strength, as one study concluded:

When rival firms issue long-term debt, their product market behavior is driven by strategic considerations that would not be present if the firms had no debt or if their debt was short term. It is shown that with limited liability, a firm's behavior in product market competition can be strongly affected by its accumulated profits. In markets where firms choose output in every period, the higher is the firm's profit in a given period, the less aggressive it will be in the subsequent periods. Thus, by issuing long-term debt rival firms may induce collusive behavior over some length of time. Furthermore, the path of equilibrium prices and the degree of price fluctuations may be entirely different depending on the structure of the firm's debt.[4]

This tendency toward market behavior is not limited to how products are priced or marketed, but can just as readily be discovered in a correlation between growth in dividends and in long-term debt. In measuring this trend, confusion may be created by use of the term "debt ratio," which can mean several different forms of calculation. In evaluating long-term debt, only one of these, the debt capitalization ratio, is meaningful. The other calculations based on the name "debt ratio" include:

1. The basic debt ratio, consisting of the relationship between liabilities and assets:

$$Total\ liabilities \div Total\ assets = DR$$

2. Debt/equity ratio, based on comparison between total liabilities and shareholders' equity:

$$Total\ liabilities \div Total\ equity = D/E$$

3. Interest coverage ratio, calculating earnings available to pay interest on debt:

$$EBIT \div I = IC$$

EBIT   earnings before interest and taxes
I   annual interest expense
IC   interest coverage

4. Cash flow to debt ratio, a ratio to test working capital availability:

$$CE \div TD = CF$$

CE cash-based earnings
TD total debt
CF cash flow ratio

5. Debt capitalization ratio, the applicable analysis of long-term debt and total capital:

$$D \div C = R$$

D long-term debt
C total capital
R debt capitalization ratio

The calculation reveals situations in which long-term debt has increased year after year as a percentage of total capital. If debt is raised substantially in order to bolster dividend payments, the long-term effect is negative. It translates to lower future earnings available to fund expansion or pay dividends.

For example, a comparison of the debt capitalization ratio for two companies reveals the differences over the long term. Table 4.1 summarizes a 10-year history of debt capitalization ratio for Coca Cola (KO) and Philip Morris (PM).

The calculation of debt capitalization ratio is applied to each company based on the formula above. For KO, the outcome of total debt divided by total capital is:

Table 4.1 10-Year debt capitalization ratio comparison—prepared by the author

Year	Debt cap ratio	
	KO	PM
2006	7.0	13.4
2007	12.9	26.0
2008	11.5	59.6
2009	16.6	68.7
2010	29.9	62.2
2011	28.7	72.9
2012	29.8	95.0
2013	35.7	126.4
2014	35.9	158.0
2015	50.2	156.1

Source S&P Stock Reports

$$28{,}639 \div 47{,}080 = 50.2$$

For PM, the same formula yields a different result:

$$25{,}250 \div 16{,}179 = 156.1$$

Both companies were carrying a similar dollar amount of debt by the end of fiscal 2015. However, with the significant difference in total capital, the ratio yields a very different result. In this comparison, both companies experienced increases in their debt capitalization ratio over 10 years, while Philip Morris moved from 13.4% up to 156.8%. (The result is always expressed as numbers without percentage signs, to one decimal place).

As of 2017, PM was yielding a dividend of 4.35%, an impressive yield. In comparison, KO was paying 3.36% annual dividend yield. However, considering the relative growth in annual long-term debt, is the higher dividend yield preferable? The answer is likely to be different for a long-term equity investor than for an options trader. The equity trader will be more likely to have concerns about long-term debt more than 100% of total capital, whereas an options trader, who does not often expect to take an equity position in the company, will tend to be more interested in total return, thus preferring the higher dividend yield in spite of the disturbing trend in long-term debt.

The analysis of debt ratio appears to condemn the PM results based on growing long-term debt, especially in comparison to the same period for KO. However, in consideration of options total return, the higher dividend yield appears to be a more important consideration. In fact, over the same period, the two price charts of these companies are remarkably similar, as shown in Fig. 4.1.

For options trading purposes, how do you interpret this paradox? Even though PM's debt history is highly negative as a fundamental indicator and KO's is far more favorable, the 10-year price history is identical, not in price but in price pattern. This may be interpreted to mean that dividend and debt trends do *not* affect stock prices over the long term; that options traders focus on dividend yield rather than on long-term equity problems likely to occur in the future; and that relatively short-term options trading is not concerned with the fundamental debt trend.

This conclusion may emphasize the point that for the purpose of fundamental analysis, the debt capitalization ratio should not be ignored; and in fact, it may provide one of several trend tests to be considered when selecting

10-Year price chart comparisons

Fig. 4.1 10-Year price chart comparisons—chart courtesy of StockCharts.com

a company for a long-term portfolio. At the same time, even though fundamental analysis is a starting point for selecting companies as potential options trading underlyings, the actual value of some fundamentals (such as the long-term debt trend) has less importance than the more immediate impact of dividend yield on an option's total return.

Dividends in Option Trading Decision—Total Return

To calculate the *total return* on an option trade, dividends are taken into account. However, this is not a simple matter of adding two separate returns together. The timing of dividends also has to be considered; and the calculation has to be annualized to make two or more positions equivalent. This annualization includes both sides, option and dividend.

Using days to annualize (versus weeks or months), the formula to annualize begins with a calculation of the initial yield. Divide the premium by the strike:

$$P \div S = R$$

P premium
S strike
R return

The strike is recommended as the basis because if the option is exercise, the strike is the exercise price.

The next step is to divide the return by the number of days remaining to expiration, and multiply that result by 365, a full year:

$$R \div H \times 365 = A$$

R return
H holding period (days)
A annualized return

For example, the premium of an option is $305 and the strike is 45 (so 100 shares = $4500). The holding period will be 148 days:

$$305 \div 4500 = 68\%$$
$$\text{and}$$
$$6.8\% \div 148 * 365 = 16.8\%$$

This formula can be reflected on an Excel spreadsheet to simplify the annualization process. Enter the following on Row 1:

Column A premium
Column B strike × 100 shares
Column C = SUM(A1/B1)
Column D holding period (days)
Column E = SUM(C1/D1) * 365

Copy columns C and E and paste to subsequent rows to enter as many annualized return calculations as you wish.

This exercise demonstrates that total return is not a matter of simply adding two net returns together. The timing of dividends based on the holding period of the option will alter the outcome, often to a considerable degree. This is further complicated by (1) methods used to calculate option returns; (2) whether an option is long, short, or part of a combination strategy; and (3) whether or not the trader owns shares in the underlying. Chapter 5 studies these issues in great detail.

For the purpose of studying options, a popular method for stock selection is the basic dividend yield. For example, a trader is considering buying shares in one of two companies. The price is approximately the same. Fundamentals of each contain similar strength levels. Are these of equal value? The tie-breaker often is dividend yield (often without consideration for the relationship between dividend growth and long-term debt growth). An investor wills elect the higher-yielding company's stock with the rationale that total return will also be higher due to the dividend itself. This ignores the variation of option premium based on the current levels if historical volatility, observing that when this attribute is high, option premium also tends to be high, and vice versa.

An illustration reveals the complexity of this decision. It makes sense from the point of view of a fundamentally-based long-term portfolio to be greatly concerned with expansion of long-term debt, regardless of current dividend yield. Looking beyond this and focusing on the analytical process of an option trader, the calculation of total return on an accurate basis forms a more reliable system for picking stocks with options trading in mind. It is quite easy to arrive at a misleading result by not considering the timing of quarterly dividends.

An example makes the point: A trader intends to buy shares of stock in one of the major oil companies, for the purpose of writing covered calls and to hedge downside risk by also trading long puts. On Friday, October 9, 2016, two companies were considered. The following valuation of the underlying and its options are all based on closing values as of that date. The two charts are shown on Fig. 4.2.

Exxon Mobil (XOM) and ConocoPhillips (COP) were both trading in a 5-point range, making historical volatility similar as of the closing date of October 9. XOM's dividend was 3.46% per year with most recent ex-dividend date on August 10; and COP's dividend was 2.25% with ex-dividend on July 21. So XOM could be expected to set up future ex-dividend dates on the tenth of November, February and May. COP would report ex-dividend dates

Two-company price chart comparisons

Fig. 4.2 Two-company price chart comparisons—chart courtesy of StockCharts.com

in the third week of the following October, January and April. These dates become important in the following calculation of total return.

Assuming 100 shares of each company were purchased as of the closing prices, XOM was priced at $86.74 and COP was worth $44.22 per share. Two possible trades are evaluated to demonstrate the methods for calculating total return. This is complex at the onset because the entire issue of calculating an option's return is itself inconsistent, even before considering dividends. For the purpose of clarity, an assumption is imposed in the following analysis: Long option returns are based on profit as a percentage of the original option purchase price. However, short positions (such as covered calls) that expire worthless are calculated as a percentage of the strike price. This is rationalized by the fact that if exercised, the strike would be the transaction price for shares. An exercised short call would require the seller to deliver 100 shares at the strike and an exercised short put would result in 100 shares put to the seller at the strike.

The two comparisons are (1) buying a long put and (2) selling a covered call. First is the long put:

> XOM: Buy one January 20, 2017 put (105 days) with strike of 85, ask price 3.95. Add $9 for cost of trading, resulting in total cost of $404. The comparison assumes a sales price netting 4.95 and a net profit of 0.91 ($91).
> COP: Buy one February 2017 put (133 days) with strike of 44, ask price 3.30. Add $9 for cost of trading, resulting in total cost of $339. In this comparison, the sales price is assumed to be 4.15, with a profit of 0.76 ($75).

This is only the starting point of the dividend analysis, and often as far as it goes. A trader calculates the return of the long put based on the strike price, annualizes the result and adds dividend yield. This is inaccurate initially because the time to expiration is different for each contract. The usual method of comparison is summarized in Table 4.2.

This version of total returns includes a flawed assumption, that a full year of dividends will be earned as part of the option's total return. Although the method of annualizing the option's return is accurate, the dividend is distorted. The option's life span does not always include a full 12 months, so this makes the initial methods inaccurate. This is not a settled put. In buying long puts while shares are also earned, an argument can be made that the annual dividend continues to be earned until shares are sold. However, to

Table 4.2 Initial annualized total return—prepared by the author

Exxon Mobil (XOM)		ConocoPhillips (COP)	
Total cost, long put	4.04	Total cost, long put	3.39
Closing sale price	4.95	Closing sale price	4.15
Net profit	0.91	Net profit	0.76
0.91 ÷ cost 4.04	22.7%	0.76 ÷ cost 3.39	22.4%
Days to expiration	105	Days to expiration	133
Annualize: 22.7 ÷ 105 × 365	78.91%	Annualize: 22.4 ÷ 133 × 365	61.47%
Add annual dividend yield	3.46%	Add annual dividend yield	2.25%
Total return, initial method	82.37%	Total return, initial method	63.72%

Table 4.3 Expanded annualized total return—prepared by the author

Exxon Mobil (XOM)		ConocoPhillips (COP)	
Total cost, long put	4.04	Total cost, long put	3.39
Closing sale price	4.95	Closing sale price	4.15
Net profit	0.91	Net profit	0.76
0.91 ÷ cost 4.04	22.7%	0.76 ÷ cost 3.39	22.4%
Days to expiration	105	Days to expiration	133
Annualize: 22.7 ÷ 105 × 365	78.91%	Annualize: 22.4 ÷ 133 × 365	61.47%
Add one dividend period: 3.46% ÷ 4	0.87%	Add two dividend periods: 2.25 ÷ 2	1.13%
Total return, initial method	79.78%	Total return, initial method	62.60%

limit the calculation of total return to the dividend only, a counter-argument is that the period should be limited to the option's life. This accomplishes a true side-by-side comparison based on the timeframe of the option. Thus, the variation of dividend yield has to be taken into account. A full year's dividend is not always earned in the life of the option; a revised comparative calculation is summarized in Table 4.3.

This analysis is based on the assumption that 100 shares of the underlying are held, meaning that dividends will be earned at the rates of 3.46% per year (XOM) and 2.25% (COP) . However, the full yield will not be earned during the option's life. The purpose in analysis of total return is to accurately compare total return on the option. In the case of XOM, the previous ex-dividend date was August 8. This means that over the ensuing 105 days, a dividend will be eared only for one quarter (November, with the option expiring in January and before the February ex-dividend). As a result, the annual dividend has to be divided by four to arrive at the true annualized return of the option.

In the case of COP, the previous ex-dividend date was July 21. This means only two quarterly dividends will be earned over the next 133 days, in October and January. The option expires in February, before the April ex-dividend date.

Annualization, for either long or short positions, is performed at the point of entry with the assumption that positions will be held until the last trading day. Realistically, a position is more likely to be closed before that date. A trader is likely to close once the percentage or dollar level of profit is realized, or once the predetermined loss has occurred. A short call can also be exercised early immediately before ex-dividend date, meaning the option will cease to exist and the trader earns a capital gain on stock (assuming the call's strike was higher than the initial price year share). In that case, the true annualized return has to be adjusted to reflect the number of days between its initial sale and the early exercise. In this case, the current dividend is not earned. This method of calculating total returns is called the *discrete method*.[5]

In a precise method, dividends expected to be earned are calculated based on a discount from the start date to ex-dividend date for each quarterly dividend, using an assumed risk-free interest rate. However, for the purpose of comparisons between potential stock and option positions, the accurately calculated annualized return is usually close enough to produce a reliable result.[6]

In these examples, return was calculated based on the original price of the put versus its profit. In the case of short options, it is more accurate to base the return on the strike of the option. The choices in calculation of overall option return is explored in detail in Chap. 5.

In order to make a true comparison between two or more choices of stocks to hold and options to buy or sell, the initial assumption has to compare outcomes to the end of the option's life, meaning the dividend to be earned in the same period has to be reflected accurately. The end of the option's life usually means last trading day, but could also end up as early exercise date.

The variation of dividend timing can be significant. It is possible to have an option open for or 89 days (24% of the year) without earning any dividend at all (if the first ex-date is on same day as purchase of shares, and the second ex-date occurs after the option expires). It is also possible to earn two quarterly dividends in a similar period, which is equal to half a year's yield. Consequently, the correct and most accurate version of total return takes two elements into account: the annualized option premium plus dividends earned during the maximum life span of the option, based on timing of shares purchased and the number of dividend earnings dates until last trading day. (For purposes of these calculations, the "earnings date" is the day before ex-dividend date, the last possible date to hold shares and earn that quarter's dividend).

The Lumpy Dividend Effect

The inclusion of dividend yield complicates the calculation of return from options trades when shares of stock are held. In fact, the Black-Scholes pricing model (see Chap. 10), excludes dividends altogether. The original formula assumes no dividend payment. Subsequent adjustments to the original formula attempt to remedy this exclusion; however, it raises the important issues that reliance on any pricing model may have to accept a series of flaws that bring into question the basic accuracy of the model.[7]

In any attempted fix to a pricing model, the effect of future dividends has to be estimated or assumed to continue at current levels per share. There are two methods in adjusting for future dividends. In a continuous dividend yield model, an assumption is made that the current dividend will continue into the future at the same rate.[8]

Second is reliance on a dividend schedule based on recent history. In this adjustment, a series of ex-dividend dates, payment dates, and dividends per share is assumed based on how recent dates and amounts occurred. The dividend yield is varied based on historical dividend behavior (raised, level, declining or skipped dividends). This is the dividend schedule method, also called application of *lumpy* dividends. The tendency for dividends to not conform to a predictable schedule introduces a variable in calculating future returns. This tendency is more closely associated with younger, lower-capitalization companies whose dividend policy is likely to adjust based on annual levels of earnings. A study of this range of companies concluded that.

> … both analytic and quantitative results show that younger firms with small capital tend to invest more in capital, and they withhold paying dividends. These firms initiate dividend payouts mainly to reduce the increasing cash holding and investment costs, as capital is accumulated.[9]

This unpredictability of a dividend trend (lumpy, or sporadic payments and timing of payments) makes it necessary to "create a list of ex-dividend dates and estimate the dividend payment for each date. This estimate is usually created from historical data and introduces a possible source of error into option pricing."[10]

A conclusion based on these points is that even with the appearance of a predictable dividend trend (yield, dividend per share, and timing of ex-dividend and payment dates), a potential variable has to be taken into account in predicting future dividend trends *and* total return on options. However, because options trades often are short-term in nature, the duration of dividend uncertainty (lumpy tendencies) is likely to not be a key issue in the calculation. Most dividend trades involving ownership of shares as part of an option strategy will be focused on short-term expirations. For long-term investment, the risk of lumpy dividend disruption to an assumption income stream is relevant as an issue of returns on equity rather than returns from trading in options.

Lumpy dividend uncertainty further complicates the calculation of cumulative return in investment. Unlike total return involving isolation of dividend yield to the period of an option's life span, cumulative return is the overall growth from inception to the current date. It often ignores annual return and reflects overall returns only. For example, an investment 10 years ago at a price of $16.50, with current value of $65.15 represents cumulative return of 394.8%:

$$\$65.15 - \$16.50 = 394.8\%$$

This represents an average of 39.48% per year over the 10-year period:

$$394.8 \div 10 = 39.48\%$$

This cumulative return does not calculate year-to-year increases in value, nor does it include dividends earned during the same period. If the average dividend was 3% per year, it adds a total of 30% (average) over the 10-year period. This is further complicated if dividends are reinvested to buy additional partial shares each quarter. This creates a compounding effect. The compounded rate is not significantly higher in a 1-year period; however, over a decade, an average of 3% compounds to 34.4% (averaging 3.44% per year) versus nominal annual interest of only 3.0%. Because compound interest grows over time, the total over 10 years will be greater than the nominal dividend yield. This simplified description further assumes that the dividend yield will remain the same over 10 years, which is not likely. Dividends may be skipped, lowered or raised, all depending on overall earnings and available cash to declare and pay dividends.

Additional Dividend Calculations

Beyond the calculation of total return, dividends further affect selection of stocks as the equity side of a dividend strategy, through a range of trends. These include the trend in dividends per share, dividend yield and payout ratio.

Dividends per share is the dollar value of annual dividends, a factor some traders consider on its own merit. However, this ignores the relative value of the dividend. For example, a $25 stock paying dividend per share of $1.00 per year yields 4%. A trader might prefer the higher dollar value of dividend per share in the amount of $8 paid on a $250 stock. However, the two investments can be made equivalent by purchasing more shares of the lower-priced stock:

$$1000 \text{ shares @ } \$25 = \$25,000 \text{ (dividend @ 4\%} = \$1000 \text{ per year)}$$

$$100 \text{ shares @ } \$250 = \$25,000 \text{ (dividend @ 3.2\%} = \$800 \text{ per year)}$$

This reveals that the annual dividend per share is far less important in the selection of one stock over another, than the dividend yield. This ultimately affects total return of an options trade as well. The dividend yield is the sum of dividend per share, divided by current price per share:

$$D \div P = Y$$

D dividend per share
P price per share
Y dividend yield

For example, the current dividend per share is $1 and price per share is $25. Yield is:

$$\$1 \div \$25 = 4.0\%$$

This brings up a related issue: Dividend yield changes every time the stock price changes. Thus, the effective dividend yield applied to every options trade is the yield based on actual purchase price of shares, and not the current yield. As price moves up, yield declines; and as price moves down yield increases. Table 4.4 demonstrates how this applies.

If the purchase price of stock was $42 per share, the dividend yield would be 4.00% as long as stock was kept, regardless of how high or low the price moved. A common error is to cite the prevailing yield when stock has declined in value. For example, recognizing a yield of 4.67% once price fell to $36 per share is inaccurate, since the original price was six points higher. The same applies as price moves upward. For options trading, calculation of total return is affected directly by this changing yield. If stock is purchased in advance of entering a trade, the effective yield should be based on the original purchase price of shares.

Table 4.4 Dividend yield—prepared by the author

Price per share ($)	Dividend per share ($)	Dividend yield (%)
36	1.68	4.67
38	1.68	4.42
40	1.68	4.20
42	1.68	4.00
44	1.68	3.82
46	1.68	3.65
48	1.68	3.50

A final calculation is the payout ratio. This is the result of dividing dividend per share by the current earnings per share (EPS):

$$D \div E = P$$

D dividend per share
E earnings per share
P payout ratio

For example, dividends for a company were most recently reported at $3.44 per share, and EPS was $4.80. The payout ratio is:

$$\$3.44 \div \$4.80 = 72\%$$

The ratio is normally reported in rounded percentage point values. For options traders intent on finding stocks with positive dividend history, the dividend yield and dividend per share are reliable trend indicators. However, the payout ratio tends to be volatile and often unpredictable. It should not be expected to increase each year, as a high ratio would not be sustainable. Some portion of earnings should be maintained for working capital management, so paying out 100% of earnings in the form of dividend would leave nothing for retirement of debt, purchase of capital assets, and business expansion.

All of the calculations involving dividends relate to calculation of total return, meaning this fundamental indicator is necessary in the selection of a stock. The apparent outcome based solely on option premium and time, but without consideration of the dividend can mislead the trader. For example, two different options each expire in the same timeframe and each yields an annualized return of 15.50%. However, one company pays a dividend of 4% and the other pays only 2%. The total return of these two examples varies from 19.50% for the first, and 17.50% for the second. This is further adjusted by the timing of ex-dividend dates.

The argument can be made that dividends should always be added at the full annual yield, based on the assumption that ownership of stock may continue even after expiration of the option. However, because the purpose in calculating total return is to estimate the *option's* total return regardless of stock price behavior, total return should consider only those quarterly dividends earned during the time between trade entry and option expiration.

This rationale is the same as treatment of capital gains on shares of stock. This should be excluded from total return even if the option were exercised and stock sold at a profit (or a loss). Because capital gains will vary considerably, the total return would be distorted if these gains were included. For example, two different stocks are held in a portfolio, each currently valued at $110 and each with identical purchase dates, dividend yields, option premium and annualized total return. However, shares of the first stock were purchase at $100 per share, and the other at $80. If both sets of stock were sold or exercised on the same day, the yield on each would be vastly different:

$$(\$110 - \$100) \div \$100 = 10.00\%$$

$$(\$110 - \$80) \div \$80 = 37.50\%$$

Adding these capital gains to total return (option profit and dividend yield) would distort the outcome is favor of the stock with lower basis. Although capital gains are true gains and cannot be ignored, they cannot be included in the specific calculation of option trade outcomes.

The next chapter delves further into the complexities of calculating returns on options trades.

Chapter Summary:

- Dividends play a complex role in the selection of options trades
- Dividend trades are best understood when trends in long-term debt are also followed
- Total return can be calculated in more than one way, but consistency is the goal
- Options expiration dates determine how many quarterly dividends will be earned
- Future dividends per share and ex-dividend dates are by no means certain
- Dividend trends directly impact total return of options trades.

Notes

1. Black, Fischer & Myron Scholes (1974, May). The effects of dividend yield and dividend policy on common stock prices and returns. *Journal of Financial Economics*, Volume 1, Issue 1. pp. 1–22.

2. Merton, Robert C. (1980, December). On estimating the expected return on the market: An exploratory investigation. *Journal of Financial Economics,* Volume 8, Issue 4, pp. 323–361.
3. Coval, Hoshua D. & Tyler Shumway (2000, June). Expected options returns. *University of Michigan Business School.* p. 2.
4. Glazer, Jacob (1994, April). The strategic effects of long-term debt in imperfect competition. *Journal of Economic Theory.* Volume 2, Issue 2. pp. 428–443.
5. Reehl, C. B. (2005). *The Mathematics of Options Trading.* New York: McGraw-Hill. p. 75.
6. Black, Fischer (1975, July/August). Fact and fantasy in the use of options. *Financial Analysts Journal,* 31. pp. 61–72.
7. Natenberg, Sheldon (1994). *Option Volatility and Pricing.* New York: McGraw-Hill. pp. 44–45.
8. Hull, John. C. (2012). *Options, Futures, and Other Derivatives,* 8th ed. Upper Saddle River NJ: Prentice Hall. pp. 269–270.
9. Nam, Changwoo (2011, August). Essays on a rational expectations model of dividend policy and stock returns. (Dissertation paper). *Office of Graduate Studies of Texas A&M University,* p. 2.
10. Chriss, Neil A. (1997). *Black-Scholes and Beyond.* New York: McGraw-Hill. p. 155.

5

Return Calculations

Chapter Objective:

- Determine a realistic net breakeven rate of return
- Compare the methods for calculating returns
- Master the various ways to calculate return on investment
- Study covered call returns as applied to the time decay strategy
- Discover the methods for annualizing net losses
- Track option outcomes to adjust net basis in stock

Total return as examined in the previous chapter adds a layer of complexity to calculation of options profit or loss. This applies when dividends are paid by the underlying company. However, the timing of ex-dividend date ultimately determines the true annualized yield from an options trade.

A problem with total return (and with any approach to calculating returns from options trades) is that a true picture is not always possible by limiting which quarterly dividends should be included. For example, a strategy in which a series of very short-term options are opened (one to two week expirations), most trades will not be made in windows of time when dividends are earned. Owning stock is a separate matter, but this timing issue complicates calculation of total return. This is one of many issues traders have to contend with, and judgment requires taking steps to ensure that like-kind comparisons are always applied to a series of trades. In addition, profitability of holding stock is not included in the option equation, but it also cannot be overlooked.

Yet another issue to contend with in return calculations is a basic question: How much return do you need to earn to accomplish a true and accurate breakeven? This is a question rarely asked by any investors or traders. The tendency is to set profit goals and bailout points without considering whether the overall outcomes are "net profitable" with all factors considered. This points to the need for calculation of a realistic breakeven point, given the impact of inflation and taxes. For many, the true net breakeven return is significantly higher than assumed.

Breakeven Rate of Return

Traders may be surprised to discover what they need to earn from trading, just to break even. Based on inflation and tax rates in effect in each situation, the breakeven rate probably is much higher than many traders realize. This is a dilemma. Taking greater risks to achieve a breakeven or better outcome also increases exposure to loss. Trading with too much caution translates to a net loss after inflation and taxes.

Options trades provide solutions to this dilemma. Numerous conservative strategies make it possible to exceed the true breakeven rate of return, notably those involving a combination of stock ownership with exceptional dividend yield, and a series of well-structured options trades (covered calls is an obvious example). Many of these involve straddles and their variations, including covered straddles and short strangles (see Chap. 8).

Breakeven analysis is applied in business decisions and product analysis, often with adjustments for time periods involved. This is acknowledged as a flawed methodology, with one analysis concluding that "the common practice of using BE analysis without considering the time value of money is likely to result in business ventures losing value instead of adding it. This can only reduce their chances of survival and growth."[1]

For analysis of options trades, the equivalent analysis of breakeven does not directly take the time value of money into account in the calculation itself. Two steps are involved: First, the necessary breakeven is calculated in consideration of taxes and inflation. Second, the element of time is brought into the analysis through annualization. The key to utilizing the analysis of breakeven is in discovering whether a stated profit goal is realistic or not, based on breakeven; and second, to determine whether attaining the required breakeven is possible within a well-defined risk tolerance level.

Traders are likely to set expectations for return on investment, usually based on a predetermined goal. For example, in a particular trade for a long option costing $100, the goal might be set to double to $200 before closing, or to accept a loss of $50. Setting a dollar amount or percentage of profit or loss is a wise starting point, even though traders tend to settle for lower profits, or let positions ride until total losses occur. The indecision and failure to execute a closing trade once goal or bail-out levels are reached, accounts for many of the losses options traders experience. The discipline required to follow goals previously set is perhaps the greatest challenge, even to experienced traders.

However, a related challenge is to determine the answers to a two-part question: First, how much net return do I need in order to achieve breakeven after inflation and taxes? Second, what options trades conform to my risk tolerance while enabling me to meet or exceed breakeven?

To calculate breakeven, two elements have to be determined. First is the rate of inflation. You can use the published Consumer Price Index (CPI) Rate published by the Bureau of Statistics (www.bls.gov/cpi) or arrive at your own estimate of inflation based on housing, transportation, utilities and food expenses.

Changes in the inflation rate tend to affect expectations of returns from stock and option trading, not to mention future dividend yield. Thus, inflation influences not only investor expectations, but also investor behavior. There is an element of irrationality in this, especially related to anticipated future dividends, as one study documented:

> … inflation raises expected nominal dividends, and on the other hand it creates a pessimistic outlook among investors, resulting in reduced expectations of future dividend payments. The direction of the overall effect is not clear, which is consistent with the fact that nominal stock returns do not increase as a result of inflation, and in many cases are even negatively affected by it.[2]

Inflation is an element in expected outcomes as well as in actual breakeven return. This means that in determining which rate to apply (CPI or a different rate based on individual perceptions), the resulting breakeven will vary as well.

In addition to calculation of inflation as it applies to trading activity, the second element is the effective tax rate. Variables apply in arriving at this value, however. Different forms of investment are taxed at different rates. For example, long-term capital gains and dividends are not necessarily taxed at ordinary tax rates:

> An investor, in choosing between alternative investments of comparable risk, will normally choose the investment that is expected to yield the highest

after-tax return. To choose which investment is expected to yield the highest after-tax return, the investor must consider both the expected before-tax return on each investment and the rate at which each investment alternative is likely to be taxed.[3]

Because this is a variable, the most efficient and accurate method to use is to base the rate on the latest reported income tax rate, and assume this is the average effective tax rate required to get to a breakeven yield on trading and investment activity. The applicable effective tax rate is the rate you are assessed for taxes based on your taxable income (*TI*). Taxable income is the net of total income (*I*) less adjustments (*A*), less itemized or standard deductions (*D*), less exemptions (*E*):

$$I - A - D - E = T$$

I   total income
A   adjustments
D   deductions
E   exemptions
T   taxable income

This applies only to your federal tax return. To calculate effective tax rate (ETR), divide taxable income (*T*) by your total income tax liability (*L*):

$$T \div L = E$$

T   taxable income
L   tax liability
E   effective tax rate

Add to this the effective tax rate based on your state (and local) income tax liability. This calculation varies by state.

Once you have determined your overall federal and state effective tax rate (ETR), the breakeven rate (BR) can be calculated. This is the rate of return from trading and investing that you need to achieve after inflation (*I*) and taxes. The calculation is:

$$I \div (100 - E) = B$$

I   rate of inflation
E   effective tax rate
B   breakeven rate of return

Table 5.1 Breakeven rate of return—prepared by the author

Effective tax rate	Inflation rate					
	1%	2%	3%	4%	5%	6%
14%	1.2%	2.3%	3.5%	4.7%	5.8%	7.0%
16	1.2	2.4	3.6	4.8	6.0	7.1
18	1.2	2.4	3.7	4.9	6.1	7.3
20	1.3	2.5	3.8	5.0	6.3	7.5
22	1.3	2.6	3.8	5.1	6.4	7.7
24%	1.3%	2.6%	3.9%	5.3%	6.6%	7.9%
26	1.4	2.7	4.1	5.4	6.8	8.1
28	1.4	2.8	4.2	5.6	6.9	8.3
30	1.4	2.9	4.3	5.7	7.1	8.6
32	1.5	2.9	4.4	5.9	7.4	8.8
34%	1.5%	3.0%	4.5%	6.1%	7.6%	9.1%
36	1.6	3.1	4.7	6.3	7.8	9.4
38	1.6	3.2	4.8	6.5	8.1	9.7
40	1.7	3.3	5.0	6.7	8.3	10.0
42	1.7	3.4	5.2	6.9	8.6	10.3
44%	1.8%	3.6%	5.4%	7.1%	8.9%	10.7%
46	1.9	3.7	5.6	7.4	9.3	11.1
48	1.9	3.8	5.8	7.7	9.6	11.5
50	2.0	4.0	6.0	8.0	10.0	12.0
52	2.1	4.2	6.3	8.3	10.4	12.5

The higher your effective tax rate and the higher the rate of inflation, the higher your breakeven rate. Table 5.1 summarizes the breakeven rate for different inflation and tax rates.

If you assume a current rate of inflation of 3% and your federal and state combined effective tax rate is 46%, the table shows that you need to earn 5.6% just to break even:

$$3\% \div (100 - 46\%) = 5.6\%$$

Most stock yielding dividends do not pay at this annual rate. It might be possible to approximate this, but even at 5.6%, it is only breakeven. Your purchasing power does not grow at all if you earn this rate. Fortunately, many covered calls and similar strategies can exceed this breakeven rate consistently, making options trading one of the few low-risk methods for beating the post-tax and post-inflation breakeven rate.

As long as the trading focus is on short-term covered calls with strikes above the basis in stock, a net profitable outcome is easily achieved. The major risk with covered calls is the market risk of a stock price declining below the net basis (N). This is calculated as your original stock basis (B), reduced by the call's premium (P):

$$B - P = N$$

B basis in stock
P premium received
N net basis

For example, with a 5.6% breakeven rate as the requirement, consider the following cases, all based on closing prices on October 13, 2016:

1. Johnson & Johnson (JNJ), 2.7% dividend (ex-dividend estimated November 19, February 19, May 19 and August 19).
2. International Business Machines (IBM), 3.6% dividend (ex-dividend estimated August 8, November 8, February 8 and May 8).
3. Dow Chemical (DOW), 3.5% dividend (ex-dividend estimated December 28. March 27, June 28 and September 28).

A comparison of long-term covered call trades of these companies and their dividends demonstrates that the example breakeven of 5.6% is easily reached for all three companies. In fact, accomplishing superior net returns is not difficult for a majority of companies, especially those with attractive dividends. Table 5.2 compares these three companies and 8-month covered calls with applicable dividends. The selected expiration in these examples is 246 days away, on June 16, 2017.

In this three-company comparison, total return is adjusted to reflect only those quarterly dividends earned during the 246 days until option expirations. In all three examples, annualized return exceeds the breakeven yield requirement of 5.6%. In fact, even without annualizing option return, the three examples that all pay dividends surpass the breakeven of 5.6% (JNJ = 5.7%; IBM = 8.0%; *and* DOW = 6.0%). . Once annualized, the total return is even farther above the breakeven return.

These outcomes are not unusual. They provide examples of long-term covered calls; however, a more common practice is to write relative short-term covered calls, which creates higher annualized returns than the longer-term expirations. Shorter-term expirations yield lower dollar amounts but higher annualized yields, due to more rapid time decay. Besides time

Table 5.2 Covered call total return comparisons—prepared by the author

Johnson & Johnson (JNJ)		Int'l bus machines (IBM)		Dow chemical (DOW)	
Sell 120 call bid 4.50—trading fees	$441	Sell 155 call bid 8.30—trading fees	$821	Sell 55 call bid 2.42—trading fees	$233
Return 4.41 ÷ 120	3.7%	Return 8.21 ÷ 155	5.3%	Return 2.33 ÷ 55	4.2%
dividend	2.7%	dividend	3.6%	Dividend	3.5%
3 dividend periods, Nov, Feb and May	2.0%	3 dividend periods, Nov, Feb and May	2.7%	2 dividend periods, Dec and Mar	1.8%
Annualized: 3.7% ÷ 246 × 365	5.5%	Annualized: 5.3% ÷ 246 × 365	7.9%	Annualized: 4.2% ÷ 246 × 365	6.2%
Total return	7.5%	Total return	10.6%	Total return	8.0%

decay, stock prices to "pin" toward the closest option strike as expiration approaches. This feature represents valuable information, and it applies solely to stocks for which options are traded:

> As noted by many active traders, option contracts that exhibit high levels of open interest tend to cause the underlying stock to migrate toward the heavily traded strike price. Pinning and other strike price effects are completely absent in stocks that do not have listed options. [4]

This directly impacts returns from options trading. It is one of many variables to consider in establishing the minimum return needed to achieve breakeven, meaning each trader needs to set up the realistic level of expected return, with the behavior of options and their underlying in mind, especially close to expiration. As desirable as short-term options are in terms of time decay, market inefficiency has to be brought into the risk analysis at the same time. Inefficiency is discovered in many forms, even the most basis. The disparity between fundamental value and market value of a company's stock is one example:

> Fundamentals matter, but it takes time for the market to recognize and fully absorb the improvement in a sector's fundamentals. When the market is not perfectly efficient, the firm's market value can differ from its fundamental value.[5]

Inefficiency is expressed in other ways, as well. For options trading, price movement creates changes in volatility, notably close to expiration. At the same time, inefficiency points to the advantages of focusing on soon-to-expire short options. This enables traders to exploit market inefficiency to great advantage.

Time decay itself is a primary driving force in the selection of short-term options as desirable in comparison to less responsive long-term contracts. The author cited above noted in the same book that on the weekend prior to expiration week, the typical option loses 32.8% of its remaining time value; and that on the Thursday before last trading day, it loses another 31.3% of remaining time value.[6]

Short-term options contain these characteristics and inefficiencies close to expiration, and they tend to work in favor of short positions such as covered calls. The tendency is for rapid and at times seemingly irrational loss of time value as expiration draws closer, notably for at-the-money contracts in the final days before expiration. Thus, establishing an expected return from options trades is made easier with this price behavior in mind. Traders during this time period are able to generate greater profits than previously for short trading. With this in mind, employing breakeven analysis to establish a minimum threshold of acceptable outcomes will work in many instances in this advantageous time window. Given the reality of both inflation and taxes and the extent to which these affect return, breakeven analysis serves as a starting point, providing more accurate guidance than traditional assumptions about expected return.

However, strategies such as covered calls vary greatly due to the selection of the underlying security, the moneyness of the strike, and time to expiration. A time decay-based strategy tends to focus on very short-term options in which time decay is maximized:

> Choosing which call to write is the most important part of any covered call strategy. Because there are so many choices, covered call strategies usually fluctuate widely from one trader to the next. Some traders opt to write long-term calls in order to eliminate the hassle of rolling their positions every month. Others choose to write significantly out-of-the-money calls in order to maximize the upside potential of their portfolio. Still others choose to write at-the-money calls, drastically limiting their upside potential but maximizing their time decay.[7]

The concept of expected return itself all too often is limited to a mathematical estimate of outcomes without first establishing desired returns or correlating returns with risks. Given the limited value of reliance on known possible outcomes and calculation of probabilities, alternative methods make sense. Ignoring the breakeven rate of return needed to at least meet the costs of inflation and taxes addresses not only the importance of understanding minimum acceptable outcomes, but goes farther. It also enables traders to better understand the relationship between return and risk, and to identify options strategies that meet and surpass expectations without adding to risk levels.

Basis for Calculation: Stock Price or Premium at Risk

The idea of holding a level of premium at risk assumes (often inaccurately) that a covered call writer appreciates the range of risk in a particular strategy. With emphasis on potential profitable returns, risk often is overlooked or the assumption prevails that strategies such as covered calls are the "sure thing" of options trading. In fact, risk assessment should be an essential part of return calculation, based on strategy selection in conformity with risk tolerance.

Calculation of likely outcomes is not merely a return comparison, but also a risk assessment. Several possible outcomes may occur, so understanding the range of risk dictates whether to focus on an exit strategy or a hold strategy:

> One form of risk management for selling options is the exit strategy. One of the simplest exit strategies for securities is selling if the given security falls by a certain percentage. The same rule can be applied to selling options. If the underlying security drops by a certain percentage, the option position is closed … In contrast to the conservative approach of closing out positions, there are also stay-the-course strategies such as double-up, covered call and the rollover. These strategies attempt to make the most of a bad situation by increasing the chances to recoup or limit any loss. However, the catch is that these strategies leave the investor open to increased risk. In a worst case scenario, a company could continue to trend lower.[8]

A point of disagreement among traders arises regarding exactly how to calculate even the most basic of returns on options trades. This also indicates that an understanding of risks in relation to risk tolerance may be poorly understood, or entirely inaccurate. To remedy this problem, the net profit from a successful trade should be summarized on a consistent basis and should also include annualizing returns. This is essential.

If one option trade is held open for only 22 days and another for 215 days, annualizing return makes both cases consistent. This further applies to a return earned over 422 days, which is longer than one full year.

Assuming these three trades all earned 2%, annualizing is the only way to make them comparable. Annualizing each of these:

$$(a)\ 2\% \div 22\ days * 365\ days = 33.2\%$$

$$(b)\ 2\% \div 215\ days * 365\ days = 3.42\%$$

$$(c)\ 2\% \div 422\ days * 365\ days = 1.7\%$$

The differences are apparent. However, even before calculating the return, what is the basis used to arrive at the initial return? Profit from an option can be based on one of three different bases:

1. <u>Original price per share.</u> If a trade is made for an option with a strike of 50, the outcome using the original price of stock will vary considerably. For example, if the purchase price was $20 per share, the outcome will be vastly different than another trade on which the basis was $48 per share. Due to the disparity in original purchase prices at the time an options trade is initiated, using the original cost per share is too inaccurate to provide consistency or even accuracy in determining return from a trade.
2. <u>Price per share at the time of the trade.</u> An alternative is to base return on the price per share when the trade was entered. This accounts for variation in the moneyness of the option and also determines the degree of expected return. If the current price per share is very close to the option's strike, expectations will be higher than if the strike is many points away from the current price per share. Using the current price is more accurate, but it makes case-to-case comparisons unreliable. If one trade consists of a 1-point gap between price and strike whereas another is set up with an 8-point gap, the outcome will distort the true return between the two cases. Depending on when shares were purchased, there are instances in which a trader will select a strike very close to the current price, and in other cases will select a strike deep in or out of the money. This makes the current price per share inaccurate.
3. <u>The strike of the option.</u> When the return is based on the strike of the option, it reflects an accurate method. First, the strike determines the value of the option based on its current moneyness. Second, the strike is selected by the trader based on the two other factors, original purchase price and

current price per share. Third, the strike is the price at which an option will be exercised unless it is closed or rolled forward prior to expiration. The strike of the option is the most reliable basis on which to determine the initial return (before annualization) of the option. This is reasonable both for long and short option trades. The only distinction in these two cases is who will determine whether or not to exercise the option.

The assumption in this comparison of three methods is that a trader is willing to accept the premise that a return can be calculated accurately based on the premium alone, regardless of how much net cash is at stake. This itself is a variable, depending on whether ownership of the underlying security is part of the trade (for example, a covered call mandates ownership of shares, whereas an uncovered call or put does not).

For options trades involving no position in stock, an alternative method for calculating return is based on initial purchase or sale of the option. For example, a long position opened for $50 and closed for $75 equals a 50% profit; and a short position opened for $100 and closed for $75 equals a 25% net profit. Losses are calculated in the same manner, based on the opening premium either received or paid. This becomes more complex if and when a short position is exercised and shares of stock put to the trader or called away. In those instances, the true net should be based on the overall net capital gain, including both stock and option outcomes.

For more precise calculations based on cash at risk versus net return, the exact calculation has to allow for variations as well as for the time capital was committed to a trade.

Return on Investment

The many ways in which return is calculated may include a very precise calculation based on cash at risk. In cases when cash is leveraged through use of margin account trading, the return should take interest expense into account as well as trading costs on both sides of the trade.

Return is derived from a trader's willingness to accept a given level of risk. Although this is an obvious conclusion for all forms of trading and investing, the desire for optimal performance may lead some to take risks beyond their defined risk tolerance:

> If all investing is risk taking, it is paramount for investors to accept only risks that are adequately rewarded. But many investors and investment managers alike fall

into the trap of predicting future returns in order to estimate which risks will be most handsomely rewarded. In most instances, investments combine both risks and opportunities, and it is virtually impossible to assess ex ante which will prevail in the future. As a result, a return estimate is always forced to represent, in one number, all the conflicting forces that affect a certain investment.[9]

Thus, calculating returns may easily distort perceptions of risk at the beginning of the process, which ultimately defeats the purpose of risk analysis. A simplified calculation that compares risk and return may enlighten a trader to a degree, but this simplified approach, notably for long options, is not always adequate to identify return on investment not only as the yield itself, but also as a factor of risk levels involved.

The return calculation is more complex when trading short options, because initial collateral and maintenance collateral are calculated as a percentage of the potential strike price. Thus, cash at risk consists of the collateral requirement, adjusted by premium received. This is easily distorted, however, if a trader has adequate cash in the margin account at the time a trade is opened. There is no additional payment into the account to meet the collateral requirement. This brings up a question: Should the trade be based on the collateral requirement within the margin account, or should the basis for return be limited to calculated net profits in the option trade regardless of margin? Arguments can be made in both directions, but this brings up an issue further complicating the use of collateral as a basis for calculating return.

In the case of a short, uncovered option, collateral is clearly defined by the rules in effect. However, for covered calls and for long option positions, no collateral requirements apply. For certain spreads, notably vertical spreads, collateral is limited to a calculation based on the net difference between the two strikes. There is considerable variation in collateral requirements based on the nature of the trade. With this in mind, consistency points to favorable calculation of returns based on strikes of options, whether collateral is part of the trade or not.

> SIDEBAR: **Valuable resource:** Two free resources help with margin calculations, both available from the CBOE. First is the Margin Manual, which explains precisely how margin is calculated for each type of trade: http://rtquote.cboe.com/learncenter/pdf/margin2-00.pdf
>
> Second is the margin calculator, on which the specific attributes of the trade are entered to determine the amount required to have on deposit in the margin account:
> http://www.cboe.com/tradtool/mcalc/

Confusion in the markets has been caused by reference to options collateral requirements as part of the trader's margin account. However, the use of "margin" for stock trades is considerably different than the options collateral requirements in the margin account. Stock purchasers use the margin account for leverage, paying 50% for shares of stock and borrowing the other 50% from the brokerage firm. In the case of options trades, no leverage is allowed in the payment for long options, and short options are subject to collateral requirements based on the strike and on the risk of exercise.

A further complication for calculating collateral and also for identifying the return on an options trade occurs when a position is subject to ratio treatment. Many combination trades can be adjusted to set up ratios, including ratio call spreads, calendar spreads, and calendar combinations; ratio put spreads and backspreads; and ratio application in combined call and put positions (box spreads, butterflies and condors, for example). In these instances, collateral is more complex than for a simple one-option trade and should be determined using the margin calculator. Return for a complex ratio strategy is likely to be determined by overall performance of the position, which can involve closing in legs rather than in one trade. This complicates not only overall return, but the method for annualizing each leg in the trade.

The Rate of Return Calculation

Given the variations in option positions, the basic rate of return is calculated using the same formula (even though it might need to be applied to various legs of a complex trade). The formula for return is:

$$(C - O) \div S = R$$

C closing value
O opening price
S strike
R rate of return

This is applicable to both long and short trades and for consistency, should represent the difference between the original trade and the closing or expiration value. Thus, an option expiring worthless has maximum difference between open and close and is advantageous for short sellers (or used as the basis to calculate the net loss for long options).

In a short trade, the closing value is equal to the "buy to close" price adjusted by trading costs; and the opening value is equal to the "sell to open." These are opposite of the long position's "sell to close" and "buy to open" trades.

The calculation of return is considerably different than the equivalent calculation for trades in shares of stock:

$$(C - O) \div O = R$$

C   closing value
O   opening price
R   rate of return

In this trade, there is to separate strike involved, only valuation of stock. Thus, buying 100 shares at $45 and then selling at $50 is calculated as:

$$(\$5000 - \$4500) \div \$4500 = 11.1\%$$

In both of these calculations, side-by-side returns are made truly comparable by annualizing the outcome. In this adjustment, the initial "return" is transferred to "rate of return." Thus, for an options trade, the annualized rate of return is:

$$((C - O) \div S) \div H * 365 = A$$

C   closing value
O   opening price
S   strike
H   holding period (days)
A   annualized rate of return

For stock trades, annualized rate of return is calculated as:

$$((C - O) \div O) \div H * 365 = A$$

C   closing value
O   opening price
H   holding period (days)
A   annualized rate of return

In both of these calculations, the initial return is based on the differences between opening and closing values. However, it makes no allowance for annual returns, which is why annualization is essential. Annualized return for both options and stocks is more complex when options are also earned during the holding period, as described in Chap. 4.

Return on Covered Calls Trades

Return calculations for single option trades is straightforward, once the correct base is identified. For consistency, the strike price is the most logical choice. For covered calls, the same observation applies. However, two separate calculations are required. Using these enables you to determine before opening a trade what occurs in one of two outcomes: expiration or assignment.

In practice, many covered calls (perhaps a majority) will be closed well before expiration. This occurs because covered calls are designed to take advantage of time decay. Selecting expiration within two weeks or less ensures a greater likelihood of profits, so the call can be closed early to generate that outcome. A popular trading system combines closing before expiration and then replacement with a later-expiring short call with more time value. This not only generates greater profits; it also gives the trader an opportunity to adjust the strike based on price movement of the underlying.

This strategy—short-term call writing and time decay replacement—works most effectively for very short-term expirations. For those issues offering weekly calls, the positions can be rolled from week to week. For those with monthly calls only, the strategy is less flexible and mandates a longer holding period.

Other than the popular strategy of opening covered calls and a close-and-replace approach before expiration, the remaining two outcomes are based on keeping the short call open until expiration. Only about 10% of all options are exercised (due to the ability to close or roll forward, meaning exercise is more often avoidable than inevitable). In fact, 55–60% of all options are closed well before expiration and only 30–35% expire worthless. The often cited statistics are incorrect. These include common claims such as "between 75 and 90% of all options expire worthless." This high worthless expiration value applies only to options held open all the way to expiration.[10]

Closing and replacing covered calls before expiration often is a profitable strategy. The rapid decline in time value takes the call's value down far enough that it can be closed at a profit. At that same time, the higher time value in the next expiration cycle provides profitable replacement. In comparison, waiting for the current covered call to expire will yield relatively little additional profit, whereas the time value on the later-expiring option may have decayed farther than the remaining value in the current call.

The two calculations for covered calls held until expiration are *return if unchanged* and *return if exercised*. The calculation of return if unchanged assumes worthless expiration, meaning all of the option premium is profit. Based on the initial credit, this is a 100% yield. However, the more accurate calculation is based on the option's strike.

Some versions of *return if unchanged* inaccurately describe either the option premium or the value of the underlying as unchanged. In fact, these assumptions are not especially useful. If intrinsic value increases to the same degree that time value declines, the option could in theory remain unchanged. If the stock price remains unchanged, and assuming the option was opened at or out of the money, then it will expire worthless. In the case of offsets between growing intrinsic value and falling time value, holding an option open would be illogical, since the call would be in the money as expiration approaches. To avoid exercise, a trader would be motivated to close and roll unless exercise was considered a desirable outcome.

If the underlying price is unchanged or declines, the covered call remains out of the money. In that case, the option expires worthless and the return if unchanged is calculated based on the original premium of the sold option, as a percentage of the option's strike:

$$P \div S = R$$

P   premium of the option
S   strike
R   return if unchanged

As with all options returns, this should next be annualized to arrive at an accurate comparison yield between two or more covered calls:

$$(P \div S) \div D * 365 = A$$

P   premium of the option
S   strike
D   holding period (in days)
A   annualized return if unchanged

For example, a covered call is opened with a strike of 50 and the premium, net of trading costs is $315. The return if unchanged is 0.63% (3.15 ÷ 50). However, if the position was open for only 35 days, the annualized return if unchanged is substantially higher:

$$(3.15 \div 50) \div 35 * 365 = 65.7\%$$

The calculation of return if unchanged, based on expiration, is one of the two possible outcomes from a covered call trade held until it expires worthless. The alternative calculation, *return if exercised*, assumes that the option is exercised on the last trading day. An initial point about this calculation is that the option, in isolation, yields the same return as in the previous calculation. In the example, return if unchanged was annualized at 65.7%. However, because the call is exercised, it becomes part of a capital gain in the underlying.

For example, the underlying was purchased at $47 per share. The covered call was opened with a 50 strike and 35 days to go until expiration. In this case, the overall return with exercise consists of:

$$(S - B + P) \div B = R$$

S   strike for 100 shares
B   basis in the underlying for 100 shares
P   premium of the covered call
R   return if exercised

Applied to the example, in which 100 shares were bought at $47 per shares and exercised at $50 per share:

$$(\$5000 - \$4700 + \$315) \div \$4700 = 13.1\%$$

In this calculation, the capital gain on the underlying is added to the option premium to calculate return if exercised. In most option return calculations, capital gains on the underlying are excluded, but in the case of exercise, the option premium adjusts the stock's net capital gain. However, a new problem arises as a result of this. If the holding period of stock is different than the holding period of the option, the annualization has to be separated into two parts. For example, the option was open for 35 days. However, if the stock was held for 165 days, then the overall annualized return contains two parts:

$$(P \div S) \div D * 365 = A_o$$
$$((S_{100} - B) \div B)D * 365 = A_u$$
$$A_o + A_u = A_t$$

P   premium of the option
S   strike

D holding period (in days)
A_o annualized return if unchanged, option
A_u annualized return, underlying
A_t annualized return, total
S basis in the underlying for 100 shares
S_{100} strike for 100 shares

Applied to the example:

$$(3.15 \div 50) \div 35 * 365 = 65.7\%$$
$$(($5000 - $4700) \div $4700) \div 165 * 365 = 14.1\%$$
$$65.7\% + 14.1\% = 79.8\%$$

This level of complexity is necessary only when the holding period of stock and option are not identical. In practice, options traders are likely to use *return if unchanged* whether the option expires worthless or is exercised; and to calculate annualized return on stock as a separate matter.

This is further complicated when calculating total return, involving dividend yield as well as option returns. As demonstrated in Chap. 4, an accurate summary of dividend yield depends on dividends earned during the holding period of the option, which can mean a zero dividend or several partial-year dividends.

Because a majority of options are closed prior to expiration, these calculations ignore the practical advantages of closing once time value has been depleted, and replacing a current option with a later-expiring one—perhaps at a different strike. In these cases, the return is easily calculated by dividing the net profit on a trade by the strike. For example, the strike is 50 and a covered call is sold for a net premium of 3.15 ($315). Four days before expiration (meaning the option was open for 31 days), the call had declined to a value of 0.45 ($45) net of trading costs. The annualized return in this trade is calculated as:

$$((3.15 - 0.45) \div 50) \div 31 * 365 = 63.6\%$$

This outcome is slightly less profitable than the calculation of the option held all the way to expiration (35 days). However, the advantage to closing and replacing is that the replacement option will contain higher time value, so the annualized return can be repeated many times while generating recurring net credits.

These annualized returns should not be viewed as what a trader may expect to earn in a portfolio of options trading. It is a comparative device intended to allow accuracy between several options trades, held for varying periods of time. For example, holding the contract open until expiration yields 65.7% whereas closing it four days earlier for a profit of $270 ($315–$45), yields 63.6% annualized return. However, at the same time, the early close allows replacement for higher time value with a later-expiring option. If the difference was much greater, early closure would make less sense. The comparison of annualized yields is the means for determining whether the replacement and rolling strategy makes sense.

Calculating Covered Call Losses

When losses are realized on covered call trades, it must be the result of a trader taking some action. This is either closing the position at a loss to avoid exercise, or rolling forward. In a forward roll, you might tend to think of the trade as one multi-leg trade but in fact, the initial trade and the new trade are separate; however, the basis in the new trade has to be adjusted.

Example: You sold a covered call on a 50 strike and received 3.15 ($315) net of trading costs. The option was set to expire in 31 days. However, in 24 days, the call had moved in the money and in spite of time decay, the premium rose to an ask price of 3.75 ($375).

In determining profit or loss, it is important to recall that a "sell to open" is based on the bid price, minus trading costs; and that a "buy to close" has to be based on the ask price, plus trading costs:

$$\text{Sell to open}: B - C = N$$
$$\text{Buy to close}: A + C = T$$

- B bid price
- A ask price
- C trading costs
- N net received
- T total paid

In the above example, the loss on the original call, based on the strike, is:

$$(3.15 - 3.75) \div 50 = -1.2\%$$

Annualized:

$$((3.15 - 3.75) \div 50) \div 24 * 365 = -18.3\%$$

Some traders prefer limiting the analysis of loss to the exchange of values in the call premium alone. In this case, the calculation is quite different, based on the original value of the short call:
$$(3.15 - 3.75) \div 3.75 = -16.0\%$$

Annualized:
$$((3.15 - 3.75) \div 3.75) \div 24 * 365 = -243.3\%$$

Clearly, recognizing a $60 as equivalent to a 3-digit loss is one consequence of annualizing. However, the purpose of this is to make accurate comparisons between two or more trades, and not to conclude a true annualized rate of profit or loss.

As with all covered call returns, the net loss would be modified by inclusion of any dividends earned during the holding period. When large annualized losses occur, the concept of total return is easily lost in the mix. This does not mean the calculation should be changed if dividends are part of this calculation, but it is a factor worth keeping in mind in the total return calculation.

In this example, the option was rolled forward to avoid exercise. Taking the loss on the current option and offsetting that with one expiring a full month later is likely to yield a net credit in the overall trade. For example, a 50-day covered call at the same strike may be sold to open at 2.03. Deducting trading costs, the net received is 1.94, or $194:

$$\$203 - \$9 = \$194$$

The previous net loss was $60. The basis in the new call has to be adjusted downward to absorb this loss:

$$\$194 - \$60 = \$134$$

The adjusted basis in the newly rolled covered call is reduced to $134. To close at a profit, the call's premium must decline below this level, net of trading fees. This analysis points out two potential problems with rolling forward. First, the time of exposure in the open short call is extended as a result of rolling, often for a very small net credit. Second, the basis in the new call has to be adjusted so that the overall trade (loss on the original covered call deducted from basis in the new call) means that getting to breakeven or net profit is more challenging.

Overlooking the need to make the adjustment ignores the reality of the true net credit and potential profit. However, it makes sense in some circumstances to view a forward roll as the combination of a loss on one trade, and setting up another separate trade with a later expiration. Accepting small losses rather than ignoring the disadvantages in the roll is not a constructive trading policy. Once a current covered call is closed at a small loss, you may consider a higher strike for a new position, based on the fact that the underlying price has moved so that the original strike is in the money.

Calculating Adjusted Basis in Stock for Covered Call Trades

In completing a covered call trade, the net profit or loss may also be taken as an adjustment to the net basis in the underlying. For example, if you buy 100 shares at $49.50 per share and earn $60 on a completed covered call, the stock's adjusted basis is:

$$\$49.50 - \$60 = \$48.90$$

The value in doing this is in adjusting the level at which a later covered call can be opened without setting up a net capital loss. Before the covered call profits were realized, opening a covered call below the cost of $49.50 would have set up a potential capital loss. However, with the adjusted basis of $48.90, a new covered call with a 49 strike would set up breakeven or small net capital gain in the event of exercise.

The same adjustment has to be made for losses on covered call trades. For example, if you buy stock at $49.50 and sell a covered call with a 50 strike, exercise would set up a small net capital gain. However, if you closed the call to avoid exercise and accepted a net loss on the trade of $75, your adjusted basis in the underlying changes to:

$$\$49.50 + \$75 = \$50.25$$

Now, with the basis in the underlying higher than the original purchase price, any subsequent covered calls should be opened at or above the adjusted basis of $50.25.

Uncovered Option Returns

Calculating profit or loss on uncovered options is a different matter than for covered calls. For uncovered calls, the risk exposure is exceptionally high. If the underlying price moves far above the call's strike, it represents a loss in intrinsic value. You have to close at a loss, accept exercise and pay for the net difference between strike and price, or roll forward to avoid exercise. The loss on such a trade is calculated as:

$$C - S - P = L$$

- C current price at the time of assignment
- S strike of the uncovered call
- P premium received for the call
- L uncovered call loss

For a put, similar calculations apply. Because a short put can be covered only by an offsetting long put, the calculation of a stand-alone short put should be based on comparison between premium and strike. Alternatively, a return may be based on the profit or loss as a percentage of the original net sales price of the put. In both cases, the process is the same as that for the covered call, but the risk levels are not always the same.

For an uncovered call, risks are exceptionally high, compared to exceptionally low risks for covered calls. For uncovered puts, the market risk is identical to the market risk of the covered call, with some notable qualifiers:

1. Uncovered puts are not modified by dividend yield. Covered calls include stock ownership, so total return applies due to dividend yield.
2. With covered calls, the market risk in the stock is a factor, but there is no equity position for the uncovered put.
3. Covered calls may result in a net loss, when the underlying price falls below the net basis (original price paid, minus premium received for the call). With an uncovered put, the equity risk is not a consideration.
4. In the event of a net loss, a new covered call might not be available without also setting up potential net capital losses. For the uncovered put, the strike used is not a consideration, since to capital gain or loss in an underlying is involved.

Annualizing a Stock's Return

A final version of return calculation calls for an annualized rate for the underlying. Calculating this for options whose lifespan is less than one year is well understood; and total return combines options profits with dividend yield on an annualized basis. However, if you also hold shares of stock for an extended period of time, the profit or loss on that portion of your portfolio also has to be calculated to gain a broad view of outcomes.

The calculation is the same annualization process as that used for options return. However, if the stock is held for a long period of time, annualization means reducing the overall yield to reflect a one-year result, putting return on the same basis as annualized total return of the options. The formula for annualizing stock profits is:

$$((E - B) \div B) \div H * 365 = A$$

E ending value
B beginning value
H holding period in days
A annualized return

For example, you have owned 100 shares of J.P. Morgan Chase (JPM) for 1092 days, just under three years, and have written a series of options against the stock during that period of time. The annualized total return has been calculated to include options net profits and dividend yield. During this time, the stock has also appreciated from $48.25 to $67.52. So in addition to total return from options (and dividends), you want to also calculate the compounded annual growth rate on the stock.

The chart in Fig. 5.1 summarizes the growth in price per share over 3 years. Annualized stock return is:

$$(($67.52 - $48.25) \div $48.25) \div 1092 * 365 = 13.3\%$$

In a case of decline in the stock's value, the overall loss on stock has to be annualized. For example, you bought 100 shares of ConocoPhillips (COP)

Fig. 5.1 J.P. Morgan Chase (JPM), 3 years—chart courtesy of StockCharts.com

Fig. 5.2 ConocoPhillips (COP), 3 years—chart courtesy of StockCharts.com

for $64.90 nearly 3 years ago, or 1092 days. During that period, you transacted several options returns and calculated annualized total return. However, the stock priced declined to $41.67 over that period of time, as shown in Fig. 5.2.

The annualized net loss is calculated as:

$$((\$41.67 - \$64.90) \div \$64.90) \div 1,092 * 365 = -12.0\%$$

In this example, the 3-year annualized loss was −12.0%. Thus, annualization represents the average annual loss. In the overall analysis of total return as well as profit or loss on the stock itself, the annualized net profit would be added, or the annualized net loss subtracted, from the previously calculated total return.

This exercise not only evens out the interim rise or fall in the stock price. It also reveals another attribute of options trading: The sum of total returns from options and dividends, expressed on an annualized basis, might exceed the annualized net loss on the underlying stock over the same period of time.

To make this analysis complete, you will need to discover the average of annualized total returns from options trades over the period studied, and then add the annualized net return (or subtract the annualized net loss) to arrive at the overall outcome. A more complex but exact method for calculating the cumulative annual growth rate (CAGR) is one method for determining an approximation of annual returns on stock. However, an advantage in applying the same simplified annualization formula is that this expresses all returns on the same basis.[11]

Chapter Summary:

- A post-inflation, post-tax breakeven analysis is essential to set profit goals
- Return can be calculated in several ways; consistency matters
- Return on investment relates both to stock and options trades
- Covered call returns are applicable to the time decay strategy
- Net losses, like profits, can be annualized to create consistent comparisons
- The net basis in stock should be adjusted for options profits and losses

Notes

1. Freeman, M., & Freeman, K. (1993). Considering the time value of money in breakeven analysis. *Management Accounting, 71*(1), 50.
2. Kudryavtsev, Andrey, Eval Levav & Shosh Shahrabani. (2014). Effect of inflation on nominal and real stock returns: A behavioral view. *Journal of Advanced Studies in Finance, 5*(1), 56–65.
3. Toolson, R. B. (1994). Investing in after-tax-deferred assets: A guide to determining after-tax returns. *Journal of the American Society of CLU & ChFC, 48*(5), 80.
4. Augen, Jeff (2009). *Trading Options at Expiration.* Upper Saddle River NJ: FT Press. p. 27.
5. Zhang, D. (2003). Intangible assets and stock trading strategies. *Managerial Finance, 29* (10), 38–56.
6. Ibid. p. 41.
7. Longo, M. (2006). Buying a young index: A new wrinkle on familiar strategy. *Traders Magazine,* 1.
8. Elenbaas, Tony & David Tsou. (2006, Fall). Risk management for option writers. *Futures, 35,* 22–24.
9. Klement, Joachim. (2011). Investment management is risk management-nothing more, nothing less. *The Journal of Wealth Management, 14*(3), 10–16.
10. Chicago Board Options Exchange (CBOE), http://www.cboe.com/learn-center/concepts/beyond/expiration.aspx.
11. Anson, Mark J.P., Frank J. Fabozzi and Frank J. Jones. (2010) *The Handbook of Traditional and Alternative Investment Vehicles: Investment Characteristics and Strategies.* New York: John Wiley & Sons. p. 489.

6
Strategic Payoff: The Single-Option Trade

Chapter Objectives:

- study payoff of options with awareness of numerous random variables
- compare speculative and hedging applications of long options
- analyze the risk profile for uncovered calls versus uncovered puts
- develop an understanding of risks associated with covered calls
- evaluate levels of expanded covered calls risks with ratio writes
- calculate adjusted net basis in a newly rolled option due to taking a net loss

What is the *likely* profit or loss from a trade? The calculation normally is performed based on a flawed assumption, that the position will be held open until expiration. In fact, up to 60% of all options are closed prior to expiration (see Chap. 5). However, the hold-to-expiration assumption is used in order to make consistent comparisons.

In these calculations, breakeven points are identified, again based on holding a position open until last trading day. This breakeven ignores the majority of instances in which profits or losses are realized before expiration. However, because this introduces so many variables, it adds consistency to assume closing a position on last trading day.

The calculation based on holding positions to the very end is further distorted by the fact that market volume is greater on options expiration days than on non-expiration days. This implies that distortion of underlying value (and thus, a corresponding distortion in ending option premium) may be possible for the minority of contracts held until final trading day.[1]

© The Author(s) 2017
M.C. Thomsett, *The Mathematics of Options*,
DOI 10.1007/978-3-319-56635-1_6

This potential distortion is one consideration in the evaluation of option payoff probability. This should be kept in mind in the following analyses, which are based on the broad assumption that pricing between opening of a position and the last trading day will be influenced only by changes in historical volatility, intrinsic value and time value, without any attempt to modify assumptions due to expiration-day price distortions.

The Probability of Option Payoff

In calculating breakeven (and identifying price ranges of outcomes of profit or loss), options traders suspend a realistic assessment in favor of a singular modeling system. In practice, an unknown number of random variables are involved in a range of outcomes:

> The hardest aspect about random variables to understand is that they are *idealized* models of reality. That is, we sacrifice a precise model of reality for the facility of dealing with precise mathematical objects. As such, random variables have several advantages over the real thing. For example, we can *simulate* real events an unlimited number of times.[2]

Looking beyond breakeven as an absolute outcome, it becomes possible to know with certainty that a profitable outcome must surpass that price level by expiration date. However, breakeven is one of many possible outcomes—in addition to profits of 5, 10, 15, 20, or 50%, losses at an equal variety of percentages—and in addition, the timing of outcomes—in a single day, 2, 3, 4, or 50 days, or more.

At first glance, it would appear that any option contains a 50/50 chance of profit or loss. This assumes that the moneyness of the option, time to expiration, and historical volatility are not factors, or that all of these factors contain equal weight. In practice, they all influence actual versus 50/50 outcome, and weighting one factor over another ends up as a judgment call for every trader. The task involves evaluation of trading risk based on all factors (distance of price to breakeven, moneyness of the strike, time to expiration, and volatility of the underlying security). These factors collectively define the risk in every trade. It is a risk universe far beyond the isolated breakeven price.

The uncertainty within this risk universe largely is responsible for reliance on pricing models, which invariably must be theoretical since no *current* model can predict future price activity. A theory or pricing model is not predictive, as it is often believed. The model simply calculates future volatility given present valuation. The application of probability through modeling for options payoff is related more to uniformity of assumptions (consistency) than to an attempt to definitively identify likely future price movement. This defies commonly-held beliefs. However, there is a solution: By describing and understanding levels of profit, breakeven and loss, traders can escape the uncertainty of modeling and discover a more rational means for comparing one strategy over another, or a strategy on one underlying over another.

The truth about theories (of probability, volatility, or options pricing) is that they are flawed, and they are poor alternatives the more reasonable pursuit of quantifying within the risk universe:

> In physics or engineering, a theory predicts future values. In finance, you're lucky if your model can predict the future sign correctly. So, what's the point? Models in finance, unlike those in physics, don't predict the future; mostly they relate the present value of one security to another. In science, when you say a theory is right, you mean that it's mathematically consistent and true - that is, it explains and predicts its corner of the universe. In finance, right is used to mean merely consistent: many models are right but usually none of them are true.[3]

Moving away from the theoretical and toward the practical, can two different trades be compared accurately? The comparison between two or more trades is never equal, due to the number of variables involved. The tendency to look solely at the breakeven analysis is only part of the overall risk evaluation. Among the variables at play, the greatest variable of all is the human element, the judgment and attitude of the trader. This has to be kept in mind in performing risk analysis and in analyzing the potential of a trade based on all of the variables. The human tendency among traders toward confirmation bias, stubbornness in faith of a position, or even overlooking of actual variable facts, represents the broader risks in all trading, and in particular, in options trading.

Traders seek underpriced options when going long (just as those going short look for overpriced options). A trader faces two challenges in identifying the appropriate opportunity. First is overcoming their own confirmation bias, identifying only those contracts expected to meet performance levels and expectations. Second is developing an understanding of *why* a particular option is priced at advantageous levels. These reasons may include expectations of changing volatility or observations of recent volatility trends:

There are times when most traded options seem underpriced, and times when most traded options seem overpriced. Again, there are two kinds of possible explanations for this. It may be that the market is expecting volatilities to be generally lower or generally higher than the estimates used in the formula, or it may be that factors unrelated to option values are affecting the option prices. Sometimes there should be a different volatility estimate for each option maturity. If the volatility of a stock was unusually high in the latest month, we might project a gradual decline in volatility back to more normal levels. On the other hand, if the volatility of the stock has been increasing in recent months, we might project a continued increase in volatility for a time.[4]

The assumptions traders bring to the analysis may be instinctive or based on mathematical modeling. Either may be flawed. With this in mind, a great advantage is gained by traders who are willing and able to question their own assumptions and to rely less on mathematical certainty (in breakeven analysis) and more on objectivity. This demands a willingness to question even strongly-held beliefs: "He who cannot put his thoughts on ice should not enter into the heat of dispute."[5]

Risk and Payoff Calculations: Long Calls and Puts

The most basic of payoff calculations is for the long option. This normally is an entry trade for new options traders, and is perceived as having a relatively low risk. The selection of an option is based on the low cost for options selected slightly out of the money and expiring in the near future.

Use of single long options can be a simple speculative trade, meant to take advantage of anticipated directional trends in the underlying. Considerable debate has centered on whether speculation is ethical or not, and even whether it is harmful to the broader economy. At the center of this debate is speculation in options:

> On one side of the debate, many have criticized speculation as unethical and harmful for the broader society. These critics argue that speculators increase market volatility and cause markets to be inefficient - speculators potentially cause bubbles in vital commodities and, thus, hurt the consumers, and occasionally artificially depress prices of vital commodities, thus hurting the producers. On the other side of the argument, many have stated quite the opposite - speculators improve market efficiency by usually buying when prices are below their fundamental values, and selling when prices are above

their fundamental values. Additionally, speculators improve market liquidity by trading as the counterparty to commercial hedgers, and, as such, make it possible to increase the production and consumption of vital commodities.[6]

In practice, pure speculation is not a primary force in options trading. Speculation does occur, but hedging is a more likely underlying motive among traders, with the desire to reduce equity risks in a portfolio through the strategic use of options.

Using an analytical approach to how options trading is applied, the most basic form of options—the long call or put—acts as part of a conservative hedge. To qualify as "conservative," the option should be tied directly to equity positions in a portfolio. This is significantly different than speculating with long calls or puts without a corresponding portfolio position. This connection between option and equity holding reduces risks:

> The hedging errors across the different options written on the same instrument are found to be highly correlated, especially for options whose moneyness is close. This implies that the risk of option arbitrage can be significantly reduced by combining options into portfolios… arbitrageurs by employing a portfolio approach can neutralize the hedging error induced risk on individual options that arises from infrequent rebalancing… the traditional tests of efficiency of option markets have low power because they are based on testing whether options are individually mispriced. For tests of market efficiency we recommend a more powerful portfolio approach that examines whether all the options traded on a stock are jointly efficiently priced.[7]

In avoiding errors in a portfolio-based strategy, long options solve the problem of basic market risk. Thus, a long put (insurance put) eliminates market risk below the adjusted strike (strike minus cost of the put) for as long as the put remains open; and a long call exploits momentary price swings to the bottom of a trading range, providing additional income for minimal cost (hedging the cycle itself). However, these hedging standards assume that equities are well-selected and represent positions the trader/investor wants to hold for the long term. When lower-quality stocks are held or the trader does not intend to keep shares, the hedge is not intended as a risk-reducing measure, but rather as a simple speculative offset. This distinction is supported with the argument that:

> … hedging can be justified if the hedger feels that the stocks in the portfolio have been chosen well. In these circumstances, the hedger might be very uncertain about the performance of the market as a whole, but confident that

the stocks in the portfolio will outperform the market … [when] the hedger is planning to hold a portfolio for a long period of time and requires short-term protection in an uncertain market situation.[8]

Assuming long calls and puts are employed to hedge well-selected equities within the portfolio strategy, the use of long options is more conservative than the isolated use of options to swing trade. In either case, the timing of entry and exit relies of price levels of the underlying. For example, a long call may be opened at the bottom of a price decline under several circumstances:

1. Stock is owned and the price has dropped well below basis. The long call is opened in the belief that price will rebound, in which case the option will accelerate the time needed for recovery. As stock price grows, so will intrinsic value of the call.
2. Stock is owned and the investor wants to purchase insurance to eliminate market risk below the strike of a put. When price does decline below the net "risk freeze" level, market risk is completely hedged. The freeze level is equal to the strike of the put less its net cost:

$$S - P = F$$

S strike of the put
P premium of the put
F risk freeze level

3. The timing of a trade is based on identification of clear and strong reversal signals and confirmation, increasing confidence that the price move is likely.
4. Proximity of the strike to current price is optimal. This usually means the strike is at or very close to the current price of stock. Another version of proximity relates to location of the strike in relation to resistance or support. Reversal is most likely at these edges of the current trading range.

This range of circumstances converts what often is viewed as a purely speculative trade, into a conservative trade with risk greatly hedged. Even so, the long option (often perceived as a safe and low-risk trade) contains varying levels of risk based on timing of the price pattern in the underlying, proximity to resistance and support, and strong reversal and confirmation signals.

Options traders may easily simplify perceptions of risk in binary terms, based on the likelihood of a target outcome (which may be random) and then judging results against that target:

> Risk is financial in nature. It is primarily concerned with downside deviations from the target rate of return. However, if there is a good chance of coming out better than you forecast, that is negative risk (a sweetener) which is taken into account in determining the security of an investment.[9]

In practice, the range of outcomes is potentially wide, with degrees of profit or loss always a possibility. Even with widespread perception of low risks, long options contain very specific risks and may result in net losses. A long option expiring worthless represents a 100% loss. Even though the dollar amount lost may be quite low, it is a significant loss on a percentage basis. A solution to overcoming the simplified perception of outcomes, it is useful to quantify not only the absolutes of maximum profit or loss, but also the breakeven price in any options trade.

This approach is essential in order to better understand the nature of options and perceptions of risks involved. Two factors work against long options positions. First is the unavoidable factor of time decay. As expiration approaches, decay accelerates. During the last one to two weeks, it disappears rapidly. Traders may prefer long options with more time until expiration which is the second factor. The longer time until expiration invariably translates to a higher cost, meaning overcoming both time decay and cost is more daunting for longer-term options.

The Long Call

The long call is entered in a belief that the underlying price will increase. If that occurs, movement above the call's strike creates growth in intrinsic value, of one point in the call for each point of movement in the underlying. However, as expiration approaches, it is also possible that the acceleration of time decay will match or outpace the growth in intrinsic value.

Leverage is always a factor in buying long options. However, leverage has to be understood as it relates to risk, or the offset between the advantageous leverage versus the exposure to loss:

> The advantages of buying options relate to leverage and limited risk. Leverage refers to the ratio of the underlying stock price to the option premium and

indicates the multiple of shares that can be controlled by purchasing the option vs. the stock. Limited risk means that a loss on an option purchase is limited to the premium paid.[10]

To fully appreciate the pro and con of long call trading, this relationship between risk or opportunity on one hand, and leverage on the other, has to be analyzed altogether. One potentially misleading attribute of long call trading is the high degree of leverage itself. The long call controls 100 shares of the underlying, but costs relatively little. Typically, the premium cost for at-the-money options ranges between 1.5 and 2.5% for soon-to-expire contracts. This varies based on volatility. However, a comparison between ask prices at varying moneyness and expirations reveals the status of long calls. Table 6.1 summarizes the relative long call ask prices for three expiration dates as well as for different moneyness levels for calls.

The ATM calls (143 strikes) ranges from 1.6 to 1.8% for all three expiration terms. ITM calls were between 2.4 and 3.1%; and OTM contracts were between 0.6 and 0.9%. This makes the point that the moneyness of the option affects premium levels dramatically. Although this summary does not reveal great disparity in percentage costs, longer-term options report higher costs due to greater time value.

Using the ATM November 18 call as an example, the premium was 2.29 ($229). Adding estimated trading fees of $9, total cost would be $238 for the long call. Based on this, the payoff for this contract is shown in Fig. 6.1.

Table 6.1 Long call premium comparisons, Boeing (BA)—prepared by the author

Expiration and strike[a]	Ask	%
November 11		
140 (ITM)	3.35	2.4
143 (ATM)	2.27	1.6
146 (OTM)	1.03	0.7
November 18		
140 (ITM)	4.40	3.1
143 (ATM)	2.29	1.6
146 (OTM)	0.89	0.6
November 25		
140 (ITM)	4.35	3.1
143 (ATM)	2.53	1.8
146 (OTM)	1.30	0.9

[a]Boeing's price at this time was $143.05, on November 1, 2016

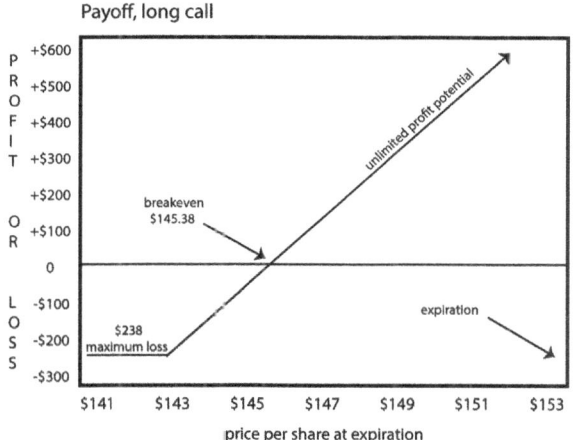

Fig. 6.1 Payoff, long call—prepared by the author

This demonstrates the attribute of a long call: Maximum loss is equal to the cost of the call, or $238; and profit potential is limited only to the extent of increase in the underlying price per share.

The visual payoff demonstrates that the maximum loss is fixed by the net premium, whereas maximum profit is unlimited. The formula for a total loss on the long call is:

$$P < S$$

P   price of the underlying
S   long call strike (net)

This maximum loss can be reduced by closing before expiration, so that a loss below 100% is realized.

Breakeven is equal to the strike plus total premium paid:

$$S + P = B$$

S   strike
P   premium paid
B   breakeven price

The premium paid in this formula has to be increased to capture the trading cost.

Profit is unlimited in theory, and it occurs whenever the underlying price exceeds the breakeven price:

$$U > B$$

U underlying price per share
B breakeven

The chart in Fig. 6.1 summarized maximum loss, breakeven and profit range in the example of the Boeing 143 option. The total cost (including trading fees) was $238. So the three possible outcomes were:

$$\text{Total loss : Underlying} < \$143.38$$
$$\text{Breakeven} : 143 + 2.38 = \$145.38$$
$$\text{Profit : Underlying} > \$145.38$$

The Long Put

A long put usually serves one of two purposes. First is the speculative expectation of price decline in the underlying, either as part of a swing trade or as a post-earnings trade. Second is the insurance put, intended to freeze market risk in the underlying at the net price of the put's strike, minus its total cost.

Like the long call, the long put is a leveraged alternative to the relatively expensive trade of shorting stock. It is also less risky, since the maximum loss in any long option is the total of premium paid. Premium cost is very low compared to an equivalent position in stock. Table 6.2 compares the long put premium for Boeing (BA) at three expiration dates and for three strikes (ATM, 3 points ITM and 3 points OTM).

The ATM puts (143 strikes) range between 2.2 and 2.5%. ITM puts were between 3.5 and 3.8%. And OTM puts ranged between 1.1 and 1.6%. This provides some idea of premium levels for long puts based on moneyness. With extended expiration, the cost should be expected to increase due to ever-higher time value premium. The disadvantage of selecting a longer-term put is its cost; the advantage is that time decay is slower than for soon-to-expire options. This adds more time for favorable (downward) movement in the underlying.

Based in the November 18 puts at 143 (ATM) strike, premium was 3.20. Adding trading fees of $9, total cost is estimated at $329 per option. The payoff for this contract is illustrated in Fig. 6.2.

The long put's maximum loss equals the premium, which in this case was $329 (including trading fees). Maximum profit is based on the extent of the move of the underlying below the strike. This is identified on the figure as

Table 6.2 Long put premium comparisons, Boeing (BA)—prepared by the author

Expiration and strike[a]	Ask	%
November 11		
140 (OTM)	1.93	1.4
143 (ATM)	2.90	2.0
146 (ITM)	5.15	3.5
November 18		
140 (OTM)	1.50	1.1
143 (ATM)	3.20	2.2
146 (ITM)	5.20	3.7
November 25		
140 (OTM)	2.23	1.6
143 (ATM)	3.55	2.5
146 (ITM)	5.60	3.8

[a]Boeing's price at this time was $143.05, on November 1, 2016

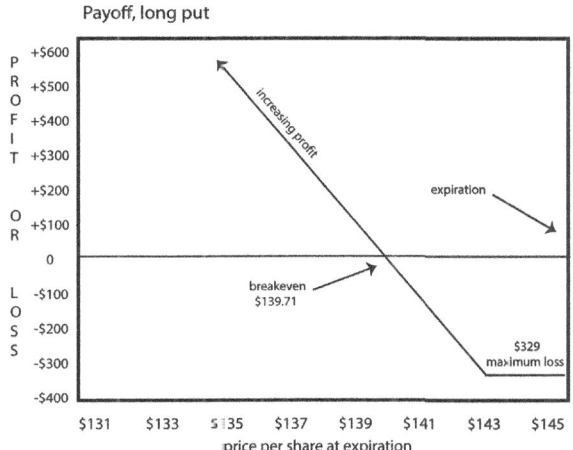

Fig. 6.2 Payoff, long put—prepared by the author

a "limited" profit potential, in comparison to the call's "unlimited" profit potential. This distinction is made because the underlying cannot decline indefinitely. The security's maximum loss is capped at a zero value of the underlying.

Maximum loss for the long put is calculated at:

$$P + F = M$$

P   premium of the long put
F   trading fees
M   maximum loss

Breakeven is equal to the strike minus the total premium paid:

$$S - (P + F) = B$$

S   strike
P   premium paid
F   trading fees
B   breakeven

Maximum profit achieved when the price is less than the cost of the put:

$$S < (C + F) = P$$

S   stock price
C   premium paid
F   trading fees
P   profit

In this case profit is equal to the net premium paid, minus the current price:

$$(P + F) - C = N$$

P   premium paid
F   trading fees
C   current price of underlying
N   net profit

The three possible outcomes for a long put in this example are[1]:

$$\text{Maximum loss} : 3.20 + 0.09 = 3.29$$
$$\text{Breakeven} : 143 - 3.29 = \$139.71$$
$$\text{Profit} : 139.71 - 129 = 10.71$$

Risk and Payoff Calculations: Uncovered Calls

Calculating payoff for uncovered options is different in many respects. Primary differences include a greater risk level and advantageous time decay. The moneyness of short positions further defines risk. A payoff analysis

---

[1] The assumed underlying price at time of expiration is $129 for the purpose of calculating profit.

restricted to potential profits from uncovered short options may easily ignore the greater risks involved.

ITM Uncovered Calls

The first uncovered option is also the highest-risk short strategy. The deeper in the money, the greater the risk. As a basic strategic consideration, a trader writing an ITM uncovered call relies on two events: a declining stock price and rapid time decay. If either of these events does not occur (or does not occur in time), the risk of loss increases.

Throughout the holding period of an uncovered call, the trader also has to be concerned with the potential for early exercise. This risk is at its height in the month of ex-dividend date. Immediately before ex-dividend, the chances of early assignment are high for ITM calls. It does not occur in every case, but this event, as part of a dividend timing strategy, points to the high risk of holding short ITM calls. By avoiding ex-dividend month, this risk is reduced considerably.

Issues such as dividends and early exercise affect the risk level of the uncovered ITM call. The calculation of profit, loss or breakeven is not an isolated quantification of the potential for the uncovered call. These factors mitigate the rationale for opening such a position. If the ITM call is deep ITM, the risk is correspondingly higher.

The position requires initial collateral and maintenance, usually equal to 20% of the strike, adjusted by the premium received for selling the call. For those intent on calculating cash-based net return, collateral is one consideration. Perhaps a greater consideration is that for such a high-risk strategy, potential profit is limited. Maximum profit is equal to the net premium received for selling the call:

$$P - F = M$$

P premium
F trading fees
M maximum profit

This profit occurs only when the underlying value is equal to or lower than the call's strike:

$$U \leq S$$

U underlying value
S strike

Breakeven is the strike of the call added to the net premium received:

$$S + P = B$$

S   strike
P   net premium received
B   breakeven

Maximum loss is unlimited for the uncovered call. It occurs when the underlying price exceeds the call's strike, minus the net premium received:

$$U - S - P = L$$

U   underlying price
S   strike price
P   net premium received
L   net loss

For example, writing deep ITM calls on Apple (AAPL), the available call strikes for three expirations are listed on Table 6.3.

The return on positions increases as the moneyness expanded farther ITM, and also as time to expiration is expanded. However, if annualized, the true picture emerges and reveals that in fact, shorter-term short options outperform longer-term short options. In this context, annualization should not be used to indicate true annual returns from short options, but does provide an explanation for the greater desirability of fast expiration terms. The closer to expiration, the more rapidly time decay works favorably for the short option. Comparing the three expirations for the 107 strikes on an annualized basis:

$$\text{November 11} : 4.1\% \div 8 \text{ days} * 365 \text{ days} = 187.1\%$$
$$\text{November 18} : 4.3\% \div 15 \text{ days} * 365 \text{ days} = 104.6\%$$
$$\text{November 25} : 4.5\% \div 22 \text{ days} * 365 \text{ days} = 74.7\%$$

Many short sellers make the mistake of viewing only the dollar value and net return of a particular ITM contract, without also considering the annualized outcome. In this view, the most advantageous option on the chart would be the one with the longest time to expiration and deepest in the money.

The three outcomes, based on the November 18 call, are:

$$\text{Limited profit: } \$465 - \$9 = \$456$$
$$\text{Breakeven: } 143 + 4.56 = \$147.56$$
$$\text{Unlimited loss: } 147.56 - 143 = 4.56$$

Table 6.3 Short call ITM comparisons, Apple (AAPL)—prepared by the author

Expiration and strike[a]	Bid	%
November 11		
109	2.82	1.3
107	4.35	4.1
105	6.05	5.8
November 18		
109	3.15	2.9
107	4.65	4.3
105	6.30	6.0
November 25		
109	3.35	3.1
107	4.80	4.5
105	6.45	6.1

[a]Apple's price at this time was $110.83, on November 3, 2016

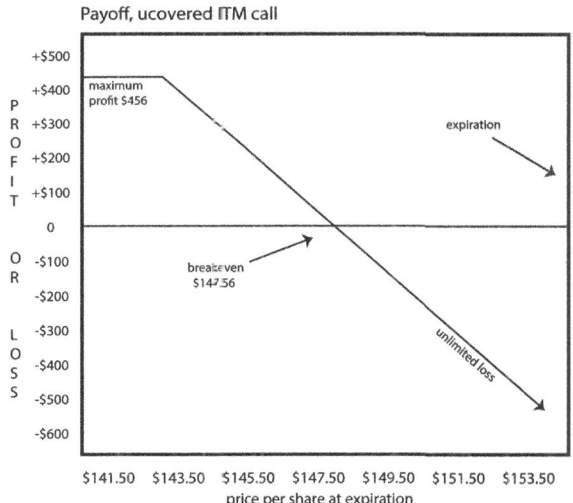

Fig. 6.3 Payoff, uncovered ITM call—prepared by the author

Using the example of the November 18 107 call, the payoff is shown in Fig. 6.3.

The illustration identified the maximum profit of $456, equal to the net credit received for selling the uncovered call. It also shows breakeven at $147.56 and the range of losses, in theory unlimited. As noted on the bottom line of the chart, the losses increase as the underlying stock price grows.

OTM Uncovered Calls

While OTM uncovered calls are high-risk trades, the risk level is somewhat reduced due to the moneyness of the option. Being OTM provides a cushion between strike and underlying price. The deeper OTM the short call, the less risk; however, at the same time, deep OTM uncovered calls also yield far less dollar value. This is especially true for soon-to-expire contracts.

The writer of any uncovered call is more likely to prefer short expirations over longer ones because time decay is accelerated. A trade-off is made between premium levels (with accelerated time decay) and level of risk.

The OTM call is less at risk for early exercise, unless the underlying price moves and the call transfers over and goes ITM. At that point, the same risk highlighted in the previous section represents a heightened risk level. Months with ex-dividend dates are the most likely timing for early exercise for ITM calls, so writing any uncovered call during ex-dividend month may be avoided due to this specific early exercise risk.

Calculation of maximum profit is the same for OTM calls as for ITM calls. It is the net of premium minus trading fees. However, as an OTM contract, the likelihood of achieving maximum profit is greater than for ITM calls. The formula:

$$P - F = M$$

P premium
F trading fees
M maximum profit

Maximum profit occurs when underlying value is equal to or lower than the call's strike. This outcome is more likely for the OTM uncovered call because the moneyness of that call is already at this price level:

$$U \leq S$$

U underlying value
S strike

Breakeven is the call's strike plus net premium received. Because the OTM call is above this level, breakeven occurs only if the underlying price rises:

$$S + P = B$$

S strike
P net premium received
B breakeven

Maximum loss is unlimited, but only occurs if the underlying price moves above the strike of the call. The farther it moves, the great the loss:

$$U - S - P = L$$

U   underlying price
S   strike price
P   net premium received
L   net loss

For example, writing deep OTM calls on Tesla (TSLA) for three expiration cycles are listed on Table 6.4.

The OTM return is calculated in the same manner as ITM calls, but with the cushion of moneyness, risks are significantly lower. The outcomes for limited profit, breakeven and unlimited loss are summarized below:

$$\text{Limited profit}: \$390 - \$9 = \$381$$
$$\text{Breakeven}: 192.50 + 3.81 = \$196.31$$
$$\text{Unlimited loss}: \text{Underlying} - 192.50 - 3.81$$

The payoff is further illustrated for the uncovered OTM call, in Fig. 6.4.

The illustration compares the limited profit of $381 maximum, to the appearance of the unlimited loss. Even though the risk is lower due to the OTM cushion, the comparison between limited profit payoff and unlimited loss exposure cannot be ignored.

Table 6.4 Short call OTM comparisons, Tesla (TSLA)—prepared by the author

Expiration and strike[a]	Bid	%
November 11		
190	3.40	1.8
192.50	2.51	1.3
195	1.77	0.9
November 18		
190	4.85	2.6
192.50	3.90	2.0
195	3.00	1.5
November 25		
190	5.40	2.8
192.50	4.35	2.3
195	3.45	1.8

[a]TSLA's price at this time was $187.42 at the close, November 3, 2016

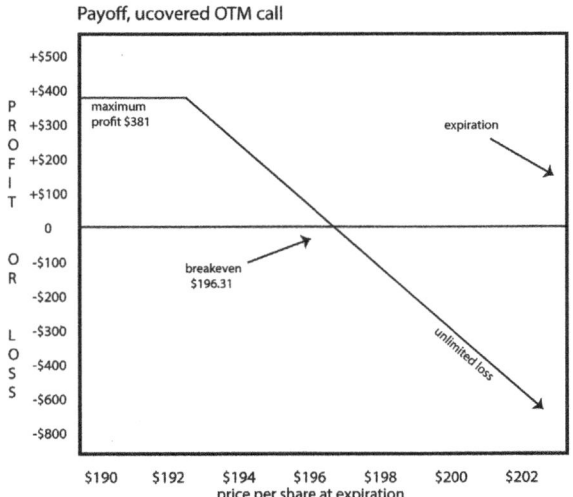

Fig. 6.4 Payoff, uncovered OTM call—prepared by the author

Risk and Payoff Calculations: Uncovered Puts

The uncovered put offers limited profit potential with downside risk. If the underlying declines below the short put's strike, the put gains value and the risk of exercise appears. However, this position can be closed or rolled to avoid exercise.

The uncovered put has the same market risk as the covered call, but without the limitations of covered calls. Specifically, if the underlying price declines below the covered call's net basis (price paid for shares, minus premium received for selling the call), recovery is possible only if the underlying rebounds or a subsequent covered call is written at or above the net strike value. Alternatively, a paper loss can also be recovered by allowing the short call to expire and then writing a short put.

Like the short call, profit potential is always limited with the short put. However, ITM short puts contain more risk than OTM due to higher exercise risk. A short put has limited profit, equal to the premium, minus trading fees. The breakeven is the net of strike, minus net premium. And the maximum loss is limited only because a stock's price cannot fall below zero. Thus, lower-strike puts contain a smaller "worst-case" outcome than those with higher strikes.

The calculation of maximum profit is the most basic and the same as the calculation for uncovered calls:

$$P - F = M$$

P   premium
F   trading fees
M   maximum profit

Breakeven is equal to the price of stock minus the net premium received:

$$S - P = B$$

S   strike
P   net premium received
B   breakeven

This is the opposite of breakeven for the short call, in which the net premium is added to the strike. A loss is calculated by subtracting the underlying price from the breakeven:

$$B - U = L$$

B   breakeven
U   underlying price
L   partial loss

The payoff outcomes for an uncovered put reply on both moneyness and time to expiration. For example, Tesla's put bids for three strikes and three expirations is summarized in Table 6.5.

The payoff diagram for the short put is opposite that of the short call. This is summarized in Fig. 6.5.

Maximum profit, breakeven and limited loss in the example are summarized based on the calculations for each:

$$\text{Limited profit} : 410 - \$9 = 401$$
$$\text{Breakeven} : 182.50 - 4.01 = \$178.49$$
$$\text{Limited loss} : 178.49 - \text{Underlying} - \text{loss}$$

Risk and Payoff Calculations: Covered Calls

The covered call might be the most misunderstood of options trades. It is widely viewed as a conservative trade, but in fact there are two risk elements often overlooked.

Table 6.5 Short put comparisons, Tesla (TSLA)—prepared by the author

Expiration and strike[a]	Bid	%
November 11		
185	3.60	1.9
182.50	2.67	1.5
180	1.98	0.1
November 18		
185	5.10	2.8
182.50	4.10	2.2
180	3.25	1.8
November 25		
185	5.75	3.1
182.50	4.80	2.6
180	3.95	2.2

[a]TSLA's price at this time was $187.42 at the close, November 3, 2016

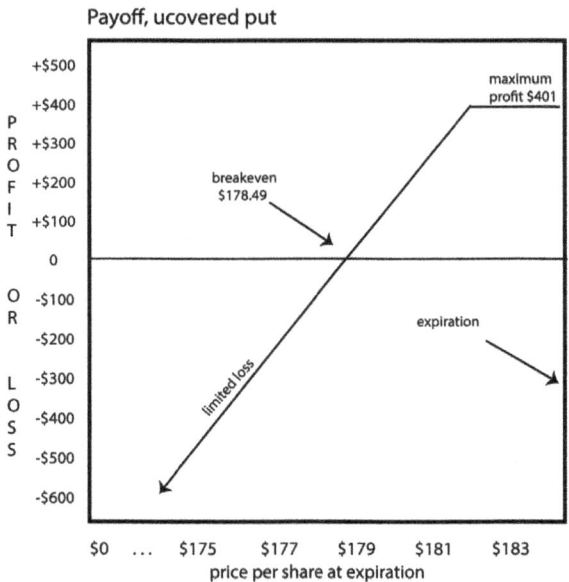

Fig. 6.5 Payoff, uncovered put—prepared by the author

The first risk element is lost opportunity. If the underlying price moves above the strike and closes with the covered call in the money, exercise results. This means shares are called away at the strike and below current market value. However, covered call writers acknowledge this risk and accept it as an outcome in a minority of instances, versus the continued earning of profits in a majority of cases.

The second risk element is related to the market price itself. If the underlying price declines, it is possible to suffer a net loss, at least on paper. For those traders unwilling to dispose of shares in this situation, there are a limited range of choices for recovery. One is to hold shares waiting for prices to return to higher levels, and to continue earning dividends. Another is to write subsequent covered calls, preferably at or above the net basis in stock. A third alternative is to write uncovered calls, with market risk identical to covered calls but without the unending concern for exercise that sets up a capital loss for depreciated stock.

Covered calls are far more conservative than many alternative trades. However, these risks exist on either side of the trade, one above the strike and cost of stock, the other below. The covered call writer needs to be aware of the breakeven prices in effect, and to be able to realistically estimate the actual risks involved with the position.

The two elements to the covered call are ownership of 100 shares of the underlying, and the sale of one call (or multiples of both stock and option based on this 1–100 relationship). The theory of the covered call is that a trader has to be willing to have stock called away at the strike. By selling a covered call, the trader grants the right to a buyer on the other side of the trade to exercise the call. This occurs only once the call moves in the money, meaning the underlying price per share moves higher thank the strike of the call.

Exercise can be avoided by closing the covered call or by rolling to forward to a later-expiring contract at the same strike or a higher strike, hopefully for a net credit in the roll itself. The benefits of being exposed to this risk are numerous. The stockholder who write a call continues to earn dividends until stock is sold or called away. The call premium is paid to the short seller immediately. Finally, the premium reduces the net basis in the underlying.

The moneyness of the covered call also defines risk, along with time to expiration. An ATM covered call will be most responsive to movement in the underlying. Many traders prefer to sell covered calls at strikes slightly out of the money, building a small cushion while setting up capital gains if and when the call is exercised. Writing ITM covered calls increased exercise risk as well as premium income. Strategically, this strategy could make sense if the call moves at or out of the money. This presents a profitable "buy to close" situation. Yet another variety of writing ITM covered calls is to set up one of two outcomes. First is exercise itself for a richer premium as well as a capital gain. Second is a rapid profit gained from the combination of time decay and movement of the underlying toward the call's strike, reducing intrinsic value and making profits easier to achieve.

Two additional events affect the risk of covered call writing and may also affect the potential for early exercise. First is the month in which dividends are earned. An ITM covered call may be exercised early prior to ex-dividend date. This does not always occur, but it is one possibility. In fact, the days prior to ex-dividend are the most likely timing for early exercise, second only to the last trading day. A trader executing a dividend capture strategy by purchasing a call a few days before ex-dividend and then exercising it on the day before ex-dividend (the last day to buy shares and earn the quarterly dividend). The shares bought with this assignment can then be sold on or after ex-dividend date. The result is that the trader earns the quarterly dividend even though the shares were owned for only one day.

From a long option point of view, this is a potentially high-profit strategy, as long as the underlying price does not decline below the net of the call's original cost versus the income from the dividend. From a short option point of view, the ITM call results in having shares called away and the loss of the quarterly dividend.

With the dividend capture strategy in mind, it makes sense to avoid selling covered calls in ex-dividend month. If a call is open and trending toward the money, it could make sense to roll forward to close the position to avoid early exercise.

The second event of concern to covered call writers (and actually to any trader with a short position, covered or uncovered) is the possibility of an earnings surprise. Earnings are announced quarterly for most companies, and any positive or negative earnings surprise is likely to have an immediate and often severe effect on a stock's price. A trader who has opened a short position may set up a vulnerable situation if the short moves in the money due to an earnings surprise. Because no one can know in advance whether an earnings surprise will be positive or negative, the short option is always at risk in this time period. It matters greatly to know when earnings announcements will be made. A prudent decision may be to roll or close a short option to avoid the day or the announcement and the days following. Avoiding this period altogether also makes sense.

In the past, the impact of earnings on stock prices tended to occur weeks before the official announcement, but a 2000 regulatory change cured this problem:

> Earnings surprises have had less impact on stock prices since the Securities and Exchange Commission passed the Fair Disclosure Regulation in 2000 …

[which] bars public companies from selectively disclosing information to certain shareholders or investors. It was passed primarily to level the playing field between institutional and individual investors - previously, companies would often disclose information on earnings in conference calls with a select group of investors.[11]

The regulatory change prevented early disclosure, but it did not reduce risk in options trading. Even though the playing field was leveled for individual investors, the immediate risk of options trading immediately before announcements may be significant. Even so, traders may view the earnings surprise as an opportunity, and might intentionally have positions open to exploit expected over-reaction in the market to even a relatively small earnings surprise. This requires perception of the likely price direction to follow earnings announcements, which also presupposes knowledge of whether the surprise will be positive or negative.

An example of how dramatically an earnings surprise affects a stock's one-day price, Table 6.6 summarizes two days of price change for amazon.com (AMZN) in late October, 2016.

On this table, price the day before earnings changed by moving up 8.56 points (prior close of $822.59 subtracted from new opening price of $831.24). However, on the day the earnings report was released, October 27, the stock declined 36.36 points, from prior close of $818.36 to new opening price of $782.00.

Anyone holding a short option faced a possible problem with this large decline. A short call, covered or not, would lose considerable value and could be closed at a profit. But a drop of this magnitude would most likely set up a net paper loss even with the discounted price per share due to the covered call. In comparison to that problem, a short put would have gained value, presenting a problem for the trader. The scope of this problem is summarized in Fig. 6.7.

One technique employed by short sellers is to select option strikes based on removal of two standard deviations from the current price. However, in this example the drop was so large that price declined below the Bollinger lower band. Price had been trading previously very close to the center band,

Table 6.6 Effect of earnings surprise, Amazon (AMZN)—prepared by the author

Date	Open	Close	Change, prior close to new open
Oct. 26	832.76	822.59	–
Oct. 27	831.24	818.36	+8.56
Oct. 28	782.00	776.32	−36.36

Amazon.com (AMZN) price chart

Fig. 6.6 Amazon.com (AMZN) price chart—chart courtesy of StockCharts.com

but the large negative earnings surprise took price lower than the normal range of trading. This demonstrates that for both short call and short put sellers, earnings surprises present an exceptionally high risk time.

An appreciate of the elements of risk (moneyness and expiration time as basic and unchanging risks, plus ex-dividend and earnings months as advanced and variable risks) helps traders to select strikes for all short options. In further articulating the overall risk, knowing the profit, breakeven and loss zones is essential for every trader.

For the covered call, capital gain or loss in the underlying is excluded from the calculation. The capital gain matters, but because it varies considerably based on the basis in stock, it is not part of the profit, breakeven or loss analysis. Maximum profit is equal to the net premium received:

$$P - F = M$$

P   premium
F   trading fees
M   maximum profit

The breakeven is calculated by treating the net premium as a discount against the strike. This, breakeven is:

$$S - P = B$$

S   strike
P   net premium received
B   breakeven

The potential loss increases as the underlying price declines. The loss is calculated as:

$$B - U = L$$

- B breakeven
- U underlying (below breakeven)
- L loss

In this case, the loss occurs as a paper loss unless shares of the underlying are sold. Although the capital gain or loss is not part of the profit or breakeven calculation, the change in paper value is used to understand potential loss levels.

Another form of loss occurs when the underlying rises above the call's strike and those shares are called away. This "lost opportunity" loss also includes the underlying value change, since the exercise of the call removes the profits that would have occurred if the call had not been sold. In this application, the call is entirely profitable but shares are called away at the strike. In order to ensure overall profitability in a covered call, the strike selected should be greater than the net basis in the underlying (original price per share, minus the call premium):

$$U - P = B$$

- U underlying purchase price per share
- P net premium received for the call
- B net basis in stock

An example of a series of call bids for OTM, ATM and ITM options and for three expirations is summarized in Table 6.7, setting up a selection of contracts for a covered call:

A covered call can be set up with any of the option contracts listed. Using the November 18 ATM 122 call with a bid of 3.40, the following analysis reveals how the covered call's outcome is calculated.

For this example, maximum profit, breakeven and loss are:

$$\text{Maximum profit} : 340 - 9 = 331$$
$$\text{Breakeven} : 122 - 3.41 = 118.59$$
$$\text{Loss} : 122 - U - 3.41 = L$$

An example of the loss: If the underlying declines to $117.50 per share, the loss is:

Table 6.7 Covered call comparisons, Netflix (NFLX)—prepared by the author

Expiration and strike[a]	Bid	%
November 11		
120	3.70	3.1
122	2.64	2.2
124	1.67	1.3
November 18		
120	4.55	3.8
122	3.40	2.8
124	2.53	2.0
November 25		
120	5.00	4.2
122	3.90	3.2
124	2.98	2.4

[a]NFLX's price at this time was $122.03 at the close, November 4, 2016

$$\$122 - \$117.50 - \$3.41 = \$1.09 \text{ per share, or } \$109 \text{ total}$$

To calculate the net basis in stock as the result of selling a covered call, the net premium is deducted from the original underlying price per share originally paid. For example, if the trader had bought shares at the price shown when the covered call was sold, or $122.03, the discounted net basis would be:

$$\$122.03 - \$3.41 = \$118.62$$

The outcomes are shown on the payoff diagram in Fig. 6.7.

Risk and Payoff Calculations: Ratio Writes and the Covered Call

Expanding the covered call into a ratio write also changes the calculation of profit, breakeven and loss. The basic ratio write involves selling more calls than are covered with underlying shares. For example, owning 200 shares of stock and selling three calls sets up a 3:2 ratio write. This may be viewed as 67% coverage (2/3rd) or as a combination of two covered calls and one uncovered call. The higher the ratio, the less risk. For example, a 4:3 ratio call write (4 calls against 300 shares) is safer than a 3:2 or 2:1.

6 Strategic Payoff: The Single-Option Trade

This strategy offers limited maximum profit:

$$(n * (P-F)) = M$$

- n number of calls
- P premium of short calls
- F trading fees
- M maximum profit

This occurs when the underlying price is equal to or lower than the strike:

$$U \leq S = \text{profit}$$

- U underlying
- S strike of the calls
- P profit

Two breakeven points apply to the ratio write, upper (when the strike is added to the maximum profit) and lower (when the strike is reduced by the maximum profit):

$$S + M = B_u$$

- S strike of calls
- M maximum profit
- B_u breakeven (upper)
- B_l breakeven (lower)

The potential loss on the ratio write is unlimited, as the underlying can rise above the short calls indefinitely. Although some of these calls are covered, the loss accumulates for the uncovered portion:

$$C - S = L$$

- C current price
- S strike price of uncovered portion
- L maximum uncovered call loss

The net loss can be avoided or mitigated by closing the call positions or rolling them forward as they transition to ITM status.

An example of these outcomes can be based on the previously introduced calls for Netflix (NFLX), on Table 6.7. For example, a ratio call write is set up based on calls expiring on November 18. Assuming ownership of 300 shares:

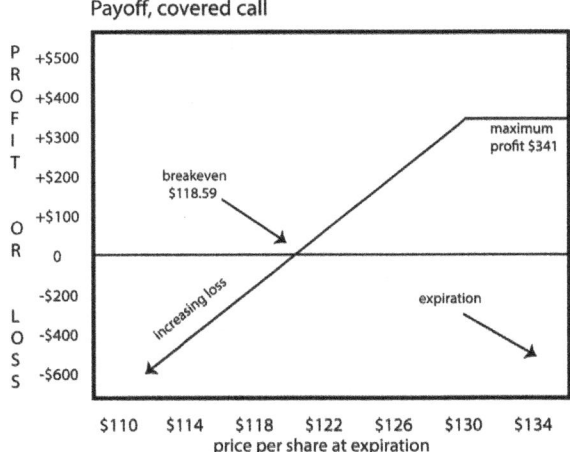

Fig. 6.7 Payoff, covered call—prepared by the author

Sell four 122 calls @ bid of 3.40, less trading fees of $11 = $1349[2]

The outcomes (for loss on options, assume current underlying price is $119):

$$\text{Maximum profit: } ((4*3.40)-0.11) = 13.49 (\$1,349)$$
$$\text{Breakeven(upper)}: 122 + 13.49 = \$135.49$$
$$\text{Breakeven(lower)}: 122 - 13.49 = \$108.51$$

Loss[3]: 125 − 122 = 3($300)

These outcomes are also represented on a diagram, as shown in Fig. 6.8.

The diagram reveals the structure of the ratio write. Maximum profit resides in the narrow area represented by total net premiums received; and potential loss occurs when the underlying moves above or below the identified breakeven price levels.

This strategy can be further expanded with the variable ratio write. In this strategy, two different strike prices are employed. This reduces overall risks. Because one strike is higher than the other, the moneyness of all options is more easily controlled through closing or rolling, if and when the underlying price approaches the lower of the two strikes.

---

[2]The fees were calculated based on the schedule with Charles Schwab & Co. The first option's trading fee is $8.75, and each additional option has a fee of 0.756. Thus, four options cost $11 to trade.
[3]assuming underlying price of $125.

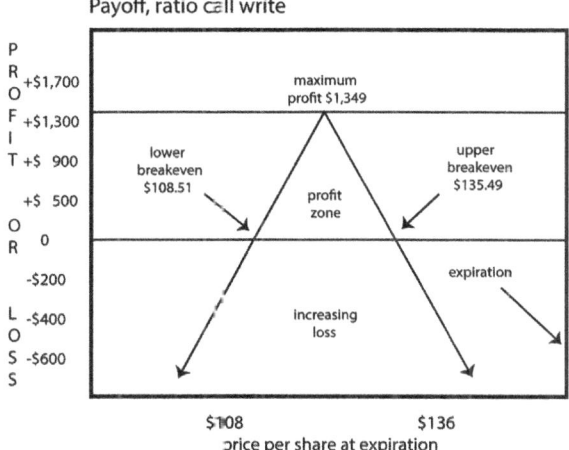

Fig. 6.8 Payoff, ratio call write—prepared by the author

Profit potential is limited to a maximum, but this is different than the previous example. With the use of a higher strike, net premium will be lower. Maximum profit is the sum of premium received, less trading fees:

$$P_{ls} + P_{hs} - F = M$$

P_{ls} premium, lower strike
P_{hs} premium, higher strike
F trading fees
M maximum profit

This occurs only when the underlying price remains at or below both strikes:

$$U \leq S_l = P$$

U underlying
S_l lower strike
P profit

The breakeven price occurs at two price levels, upper and lower:

$$S_u + M = B_u$$
$$S_l - M = B_l$$

S_u higher strike calls
S_l lower strike calls

M maximum profit
B_u breakeven (upper)
B_l breakeven (lower)

Losses occur when the underlying price moves beyond either breakeven point. The greater the extent of this move, the larger the loss. However, the advantage to the *variable* ratio writeis the higher flexibility in rolling or closing to avoid exercise. Losses on either side of breakeven occur in two instances:

$$\text{Upper price loss}: U > B_u$$

$$\text{Lower price loss}: U < B_l$$

U underlying
B_u breakeven (upper)
B_l breakeven (lower)

The calculation of each loss requires the following:

$$\text{Upper price loss}: C - B_u = L_u$$

$$\text{Lower price loss}: B_l - C = L_l$$

C Current price
B_u breakeven (upper)
B_l breakeven (lower)
L_u loss (upper)
L_l loss (lower)

An example of these outcomes can be based on the previously introduced calls for Netflix (NFLX), on Table 6.7. For example, a variable ratio call write is set up based on calls expiring on November 18. Assuming ownership of 300 shares:

> Sell two 122 calls @ bid of 3.40, less trading fees of $10 = $670
> Sell two 124 calls @ bid of 2.53, less trading fees of $10 = $496
> Total net credit = $1.166

The profit, breakeven or loss are calculated as (assuming then-current upper-level price of $135 and lower-level price of $110):

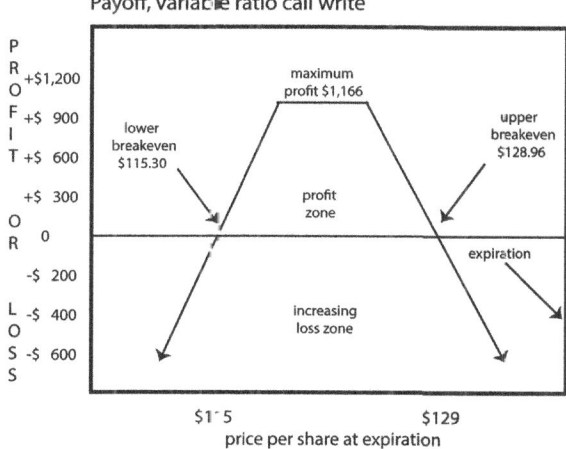

Fig. 6.9 Payoff, variable ratio call write—prepared by the author

$$\text{Profit: } 6.80 + 5.06 - 20 = 11.66 (\$1,166)$$
$$\text{Breakeven(upper): } 124 + 4.96 = 128.96$$
$$\text{Breakeven(lower): } 122 - 6.70 = 115.30$$
$$\text{Loss(upper): } 135 - 128.96 = 6.04 (\$604)$$
$$\text{Loss(lower): } 115.30 - 110 = 5.30 (\$530)$$

These outcomes are also summarized on Fig. 6.9.

The profit zone expands as the underlying price approaches either of the breakeven levels. Below and above that price, the loss zone grows.

Risk and Payoff Calculations: Covered Put

A final analysis of risk and payoff is for the covered put. By definition, a put cannot be covered in the same way as a call. However, for anyone who has shorted stock, a short call covers the risk element of that position.

In the ideal short stock trade, a trader hopes that the stock price will decline. When that occurs, the position can be bought to close at a lower price per share than the original purchase price, creating a net profit:

$$B - C = P$$

B basis in short stock (sold to open)
C current price per share (bought to close)
P profit

This scenario represents the ideal situation. However, if the underlying price rises, the short stock position suffers a loss. It is a paper loss as long as it remains open, but the risk continues as long as the underlying price rises. The transaction itself is expensive because it involves borrowing stock from the broker and paying interest on that stock as long as the short position is held. The short stock is not only risky, but expensive as well.

Short selling contains risks similar to those of a covered call. If the underlying price rises, it could continue in theory to indefinite levels. However, with a covered call, as a stockholder, a trader earns dividends. As a short seller of stock, the same trader is responsible for paying the dividend. This increases the risk.

Selling a put against short stock offsets that risk, at least partially. The two-part transaction combines shorted stock and a short put, which covers the stock position to the extent of the net premium received for the short put. Maximum profit on the short call is limited, however. It is capped at the net premium received:

$$P - F = M$$

P   premium of the short put
F   trading fees
M   maximum profit

The protection granted to the short seller is limited in the same way as a covered call. The covered put protects the short stock only to the extent of the net premium of the put. As a consequence, the covered put is of limited value, and the risk-averse short seller will have to consider closing down the short stock position once a breakeven occurs between stock and the short put. Breakeven occurs when the original "sold to open" basis price of the underlying is added to the maximum profit:

$$U + M = B$$

U   underlying price at time of short sale
M   maximum profit on the short put
B   breakeven price

The potential loss on the covered put is unlimited (in the same way as the covered call), but is increased by two factors: Dividends (which the short seller must pay) and interest expense due to the broker for lending stock to the trader.

Table 6.8 Covered put comparisons, Apple (APPL)—prepared by the author

Expiration and strike[a]	Bid	%
November 11		
110	1.28	1.2
112	2.21	2.0
114	3.50	3.1
November 18		
110	1.65	1.5
112	2.57	2.3
114	3.75	3.3
November 25		
110	1.87	1.7
112	2.77	2.5
114	3.95	3.5

[a] Apple's price at this time was $110.83, on November 3, 2016

A net loss occurs when the current underlying price is greater than the breakeven price:

$$U > B = L$$

U current underlying price
B breakeven price
L net loss

A covered put can be set up based on Apple (AAPL) bid prices of puts, as shown in Table 6.8.

Using the November 18 strike of 112 with premium of 2.57, the following possible outcomes would be realized[4]:

$$\text{Maximum profit: } 2.57 - 0.09 = 2.48 (\$248)$$
$$\text{Breakeven: } \$112 + 2.48 = \$114.48$$
$$\text{Loss}: \$117 - \$114.48 = \$2.52 (\$252)$$

These outcomes are summarized on the payoff diagram in Fig. 6.10.

The covered put demonstrates the limitations of shorting stock based on levels of risk. This position mitigates loss only to the extent of the put's premium. The preceding analysis also does not account for the cost of interest on

---

[4] The example of breakeven assumes original sales price at $112; and the example of net loss assumes the stock price rose to $117 at the time of expiration.

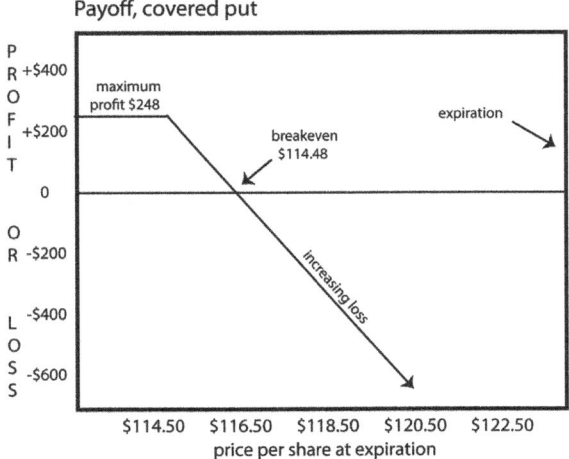

Fig. 6.10 Payoff, covered put—prepared by the author

borrowed stock, or for the cost of dividends a short seller is required to pay as long as the underlying remains open. The longer the period of time, the greater the interest cost; and if any dividend earnings occur while the short seller is stockholder of record, that is another cost factor to be considered.

Recalculation of Net Basis in Rolled Short Options

Traders selling options invariably face the prospect of exercise. The choices are to accept exercise, close the position, or roll forward. In a forward roll, the net basis in the option has to be recalculated to avoid inaccurate conclusions.

There are times when rolling makes little or no sense. Taking a small loss makes more sense when rolling increases exposure the trader does not want, or when it would be preferable to free up capital and collateral for different trades. When a short option is rolled, the formula for calculating the net basis in the new option is:

$$S_c - (B_p - S_p) = A_c$$

S_c sell to open, new option
B_p buy to close, previous option

S_p sell to open, previous option
A_c adjusted basis, new option

For example, a trader sells a covered call at a net bid price (strike minus trading fees) of 2.48 ($248). A few days before expiration, the call moved ITM and the net ask price (strike plus trading fees) was 3.50 ($350). The trader closes this position to avoid exercise and replaces it by rolling forward to the next expiration. A call with a one-point higher strike was sold to open at a net bid price of 4.15 ($415). The basis in this new call has to be reduced to absorb the loss on the previous trade:

$$4.15 - (3.50 - 2.48) = 3.13 (\$313)$$

In this example, the trader makes this adjustment to accurately reflect the net benefit of selling the covered call. By reducing the basis in the new position by the previous loss, the resulting net premium (3.13, or $313) should be used to set goals for closing the short position. If it is closed for an ask below the net of $313, a profit results. This may be the expiration price or an early close.

One problem with rolling forward is the extended distance between current premium value and breakeven. In this example, the trader sold for a bid of 4.15, but is 1.02 points away from breakeven. This will require considerable downward valuation of the short call, which will be a combination of time decay and price activity remaining ATM or OTM. This clearly presents a disadvantage for any trader selling options. The distance to breakeven is invariably increased due to the net loss. Far more desirable is a situation in which the early close of the original short option results in breakeven or a small profit. However, this is not always possible.

Chapter Summary:

- the many random variables of options and proximity complicate payoff analysis
- using long options as hedges for equity portfolios is a conservative strategy
- uncovered calls are high-risk whereas uncovered puts are low-risk
- payoff for covered calls is expanded through ratio writes and variable ratio writes
- when a short option is rolled, the replacement contract's basis has to be adjusted.

Notes

1. Stoll, Hans R. & Robert E. Whaley (1987). Program trading and expiration-day effects. *Financial Analysts Journal, 43* (2), pp. 16–28.
2. Chriss, Neil A. (1997). *Black-Scholes and Beyond.* New York: McGraw-Hill. p. 70.
3. Derman, Emanuel (2007). Sophisticated vulgarity. Risk, 20(7), 93.
4. Black, Fischer (1975). Fact and fantasy in the use of options. Financial Analysts Journal, 31(4), 36.
5. Nietzsche, Friedrich (1878). Human, All Too Human. "Man and Society," # 315.
6. Tokic, Damir (2014). Legitimate speculation versus excessive speculation. Journal of Asset Management, 15(6), 378–391.
7. Mello, Antonio S., & Henrik J. Neuhaus (1998). A portfolio approach to risk reduction in discretely rebalanced option hedges. Management Science, 44(7), 921–934.
8. Hull, John C. (2012). *Options, futures and other derivatives, 8th ed.* New York: Prentice Hall. p. 64.
9. Mao, James C. T. (1970). Survey of capital budgeting: theory and practice. *The Journal of Finance 25*, 349–360.
10. Reehl, C. B. (2003, March). Covering up with options. Futures, 32, 38–40.
11. Calio, Vince (2005). Reg FD eases effects of earnings sticker shock. Pensions & Investments, 33(7), 4–4, 45.

7

Strategic Payoff: Spreads

Chapter Objectives:

- compare differences in vertical spreads as bull or bear and using calls or puts
- adjust risk perceptions of debit and credit spreads based on profit potential
- develop an understanding of the condor and its rationale
- analyze the differences in construction of condors versus butterflies
- study the risk variables between long and short synthetic stock positions
- observe the advantages and risks of horizontal and diagonal spreads.

The spread is the most diverse of options strategies. In its most basic form, the spread consists of two options, one long and the other short. A spread may involve either calls or puts, and can be long or short. The spread can also be vertical (same expiration, different strikes), horizontal (different expirations, same strike) or diagonal (different expirations and different strikes).

Beyond spreads, many similar combinations combine calls and puts in a variety of different configurations. This chapter describes the nature of this range of strategies, and identifies the profit, loss and breakeven price levels for each.

Options traders face a challenge in trying to determine the correlation between risk and profit. The level of profit may be limited or unlimited. When limited, the question should be, Is the level of risk adequate to justify the exposure in the strategy? Likewise, the level of loss contains the same attributes, so the question should be, Is the exposure acceptable given the potential loss in the strategy?

The answer depends on whether you trade as a hedge of equity positions currently held, or as an independent trade intended to exploit information about the underlying. One study summarized the point about relative utility of options and their risks:

> When options become available, uninformed traders hedge the risk of underlying stocks by maintaining opposite positions in the options market. On the other hand, informed traders, capitalizing upon their information, hold outright positions in options.[1]

The seemingly obvious analysis of options relative to the underlying may be overlooked by traders in some cases, notably when a spread or similar strategy is perceived as a hedging mechanism for an equity position (but the trader does not hold shares). Even if the spread is set up to exploit a price pattern opportunity as part of a swing trade, the profit and loss analysis can be lost in the process and levels of exposure taken on without the accompanying risk analysis. Even experienced options traders are vulnerable to a tendency to overlook risk or to underestimate the levels of exposure that may develop while entering a trade, especially if it is then closed in legs.

Risk and Payoff Calculations: Vertical Bull Spreads

The first series of risk and payoff calculations is undertaken for vertical spreads, defined as spreads with different strikes and the same expiration. The eventual outcome relies on how far the underlying price moves in its approach to (or in excess of) the strike prices involved. So even with high profit potential, it is qualified:

> [The] bull spread options strategy ... is highly profitable when the price of the underlying primitive reaches the second out-of-the-money strike price before the expiration date of the options, but no further. The challenge lies in choosing the optimal out-of-the-money option strike price. The option exercise price, past primitive price jumps, and primitive volatility shifts are the important factors that are to be analyzed.[2]

Bull Put Credit Spread

The bull put spread sets up a net credit and combines a limited profit potential as well as limited maximum loss possibilities. It is established by selling a short ITM put and buying an OTM put at the same time and with the same expiration. The interesting attribute of using puts in a *bull* spread points out the flexibility and richness of possible options trades.

To properly determine whether profit potential versus risk exposure make this trade acceptable, you should combine analysis of not only the profit and loss potential, but also the current historical volatility levels. A trade dominated by a short position (thus setting up a net credit) is advantageous when volatility is high, and is likely to be entered in anticipation of shrinking volatility between entry date and expiration of the options. These two aspects—return and volatility—are not separate issues, but part of a holistic analysis:

> ... option trading should not be based on either volatility forecasting or return forecasting in isolation but rather as a combination of volatility and return forecasting. This combination should provide valuable information and help to improve the performance of traditional trading systems by applying advance strategies for gaining profit.[3]

For example, on November 11, Southwest Airlines (LUV) closed at $43.84. A selection of potential put strikes is summarized in Table 7.1.

The position sets up a limited profit equal to the net credit:

$$(S - F) - (B + F) = P$$

S sold put premium
F trading fees
B bought put premium
P maximum profit

The breakeven point occurs when the net premium received is subtracted from the strike of the short put:

$$S - C = B$$

S strike of the short put
C net credit received
B breakeven

Table 7.1 Calls and puts, Southwest airlines (LUV)—prepared by the author

Strike[a]	Calls		Puts	
	Bid	Ask	Bid	Ask
Dec. 16				
42	2.50	2.60	0.70	0.85
43	1.80	1.90	1.05	1.15
44	1.25	1.30	1.50	1.60
45	0.80	0.90	2.05	2.15

[a]LUV's price at this time was $43.84, at the close on November 11, 2016

Å net loss is also limited, and occurs whenever the underlying price declines below the breakeven point. The amount of loss is limited to the net difference between the two strikes, minus the net credit received:

$$S_u - S_l - C = L$$

S_u strike (upper)
S_l strike (lower)
C net credit
L maximum loss

Selling a Southwest (LUV) 44 put (ITM) at a bid of 1.50 and buying a 43 put (OTM) at an ask of 1.15 sets up a net of 0.17, or $17 net credit:

Sell 44 put, bid 1.50, less trading fees = $141
Buy 43 put, ask 1.15, plus trading fees = $124
Net credit = $17

Based on the example for LUV, the profit, breakeven a loss outcomes are: is:

Maximum profit: ($150 − $9) − ($115 + $9) = $17
Breakeven: $44 − $17 = $43.83
Maximum loss: $44 − $43 − $17 = $83

The payoff summary is illustrated in Fig. 7.1.

Bull Call Debit Spread

The second version of a bull vertical spread employs calls and sets up a net debit. Like the limited profit or loss of the put-based credit spread, the call

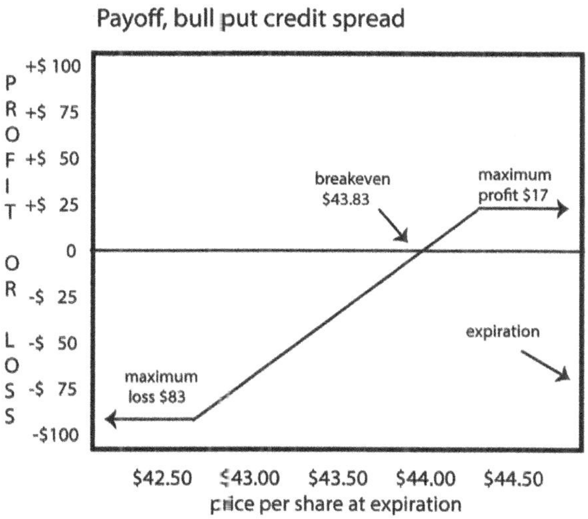

Payoff, bull put credit spread

Fig. 7.1 Payoff, bull credit spread—prepared by the author

spread's profit or loss is also limited. The difference between the two is the net credit (puts) versus net debit (calls).

Outcomes for bull call spreads are going to vary based on differences in the bid-ask spread, and for this reason an analysis of viability and of maximum profit, breakeven and loss will not be similar in every case. This spread is especially of concern in options trading:

> … the bid-ask spread does not have a substantial impact on stock investments, but it can have a dramatic impact on options trading. Therefore, to have a better idea of the return on a bull call spread, it is important to incorporate bid-ask spreads.[4]

The position combines a long ITM call with a short OTM call, both at the same expiration date. Maximum profit is realized when the underlying ends up higher than the ITM strike, and is capped at the net difference between strikes, minus the net debit:

$$(S_s - F) - (S_l + F) = P$$

S_s strike of short call
S_l strike of long call
F trading fees
P maximum profit

Breakeven resides where the long strike is added to the net debit:

$$S + D = B$$

- S strike of the long call
- D net debit paid
- B breakeven

The maximum loss is equal to the net debit paid:

$$(P_l + F) - (P_s - F) = L$$

- P_l premium, long option
- P_s premium, short option
- F trading fees
- L maximum loss

An example of the bull call debit spread, based on previously introduced Table 7.1 for Southwest Airlines (LUV) is established with the following trades:

BUY 43 call, ask 1.90, plus trading fees = $199
SELL 45 call, bid 0.80, less trading fees = $71
NET debit = $128

The outcomes for this trade are:

Maximum profit: (45 − 0.09) − (43 + 0.09) = 2.00 ($200)
Breakeven: 43 + 1.28 = 44.28 ($44.28)
Maximum loss: (1.90 + 0.09) − (0.80 − 0.09) = 1.28 ($128)

The outcomes of this trade are summarized in Fig. 7.2.

The pattern for this spread is similar to that for the bull put spread, since both are bullish versions of the vertical spread. The opposite diagram will occur for bearish spreads.

Bear Put Debit Spread

Bearish vertical spreads may consist of either puts or calls. These also limit both profit and loss potential since one side is long and the other side is short. These offsets limit risk while also limiting profits.

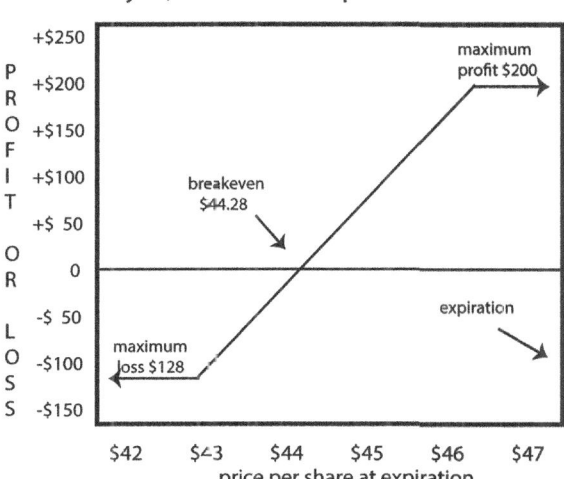

Fig. 7.2 Payoff, bull call debit spread—prepared by the author

The bear put debit spread consists of buying an ITM put and selling an OTM put with the same expiration date. This sets up limited downside profits. To generate a profit, the underlying needs to move lower, and the maximum is achieved when the underlying price is equal to or lower than the short put strike. The formula:

$$S_l - S_s - (C - F) = P$$

S_l strike, long put
S_s strike, short put
C cost of the position
F trading fees
P maximum profit

The breakeven for this strategy is equal to the long put's strike minus net premium paid:

$$S_l - (C - F) = B$$

S_l strike, long put
C cost of the position
F trading fees
B breakeven

Table 7.2 Calls and puts, Bristol-Myers Squibb (BMY)—prepared by the author

Strike[a]	Calls		Puts	
	Bid	Ask	Bid	Ask
Dec 16:				
52.50	4.30	4.50	0.51	0.55
55	2.47	2.62	1.13	1.20
57.50	1.25	1.30	2.34	2.43
60	0.55	0.60	4.05	4.35

[a]BMY's price at this time was $56.37, at the close on November 11, 2016

A loss occurs when the underlying price is equal to or greater than the long put's strike. Maximum loss is limited to the net debit paid:

$$(C - F) = L$$

C cost of the position
F trading fees
L maximum loss

For an example of how this position is constructed, Table 7.2 presents a range of December 16 expirations and strikes both in and out of the money.

A bear put debit spread is created with the following positions:

BUY 60 put, ask 4.35, plus trading fees = $444
SELL 52.50 put, bid 0.51, less trading fees = $42
Net debit = $402

The outcomes for these positions are:

Maximum profit: 60 − 52.50 − 4.02 = 3.48 ($348)
Breakeven: 60 − 4.02 = 44.98 ($55.98)
Maximum loss: (4.35 + 0.09) − (51 − 0.09) = 4.02 ($402)

This is summarized on the diagram in Fig. 7.3.

Bear Call Credit Spread

A bear spread consisting of calls will set up a net credit. Both maximum profit and loss are limited in this position due to the long-short offset in each position.

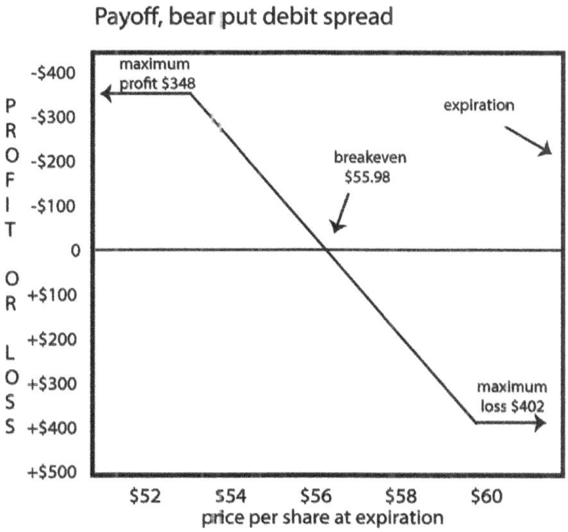

Payoff, bear put debit spread

[Chart showing payoff for bear put debit spread with:
- Y-axis: PROFIT OR LOSS from -$400 to +$500
- X-axis: price per share at expiration from $52 to $60
- maximum profit $348
- breakeven $55.98
- maximum loss $402
- expiration line sloping down from upper left to lower right]

Fig. 7.3 Payoff, bear put debit spread—prepared by the author

This spread combines a long call at an OTM strike, with the same number of short call with an ITM strike. Although any vertical spread can be moved entirely in or out of the money, this configuration—on either side of the current underlying value—aptly illustrates the possible outcomes of such a position.

As a bear spread, profit is expected if and when the underlying price declines. The maximum profit is equal to the net credit received for opening this position, and resides when the underlying is equal to or less than the strike of the short call:

$$P - F = M$$

- P premium received
- F trading fees
- M maximum profit

Breakeven occurs when the short call's strike is added to the net premium received:

$$S_s + M = B$$

- S_s strike, short call
- M maximum profit
- B breakeven

A maximum net loss will result when the underlying moves higher than the long call's strike, and is fixed at a maximum of the net differences between the two strikes, minus net premium received:

$$S_l - S_s - (P - F) = L$$

S_l strike, long call
S_s strike, short call
P premium received
F trading fees
L maximum loss

Referring to the previously introduced Table 7.2, an example of the bear call credit spread for Bristol-Myers Squibb may consist of:

Buy 57.50 call, ask 1.30, plus trading fees = $139
Sell 55 call, bid 2.47, less trading fees = $238
Net credit = $99

The net outcomes for this trade would be:

Maximum profit: ($247 − $9) − ($1.30 + $9) = $99
Breakeven: 55 + 0.99 = 55.99 ($55.99)
Maximum loss: 57.50 – 55 – 0.99 = 1.51 ($151)

These outcomes are visually summarized in Fig. 7.4.

The pattern for this call bear spread has the same design as the put bear spread. However, the profit and loss levels have to be adjusted based on the differences between strikes and premium values.

Among the many expansions of the vertical spreads is the box spread, which in its basic configuration, combines a bull call spread and a bear put spread. Although described as an arbitrage "riskless" trade, the box spread is complex and involves trading fees to the extent that potential profits are extremely limited. The box spread is not riskless in spite of the frequently used term. There is the risk of exercise on the short side:

> Assignment of a short put will incur large carrying costs on the resulting long stock; assignment of a short call will inevitably come just before an ex-dividend date, costing the arbitrageur the amount of the dividend.[5]

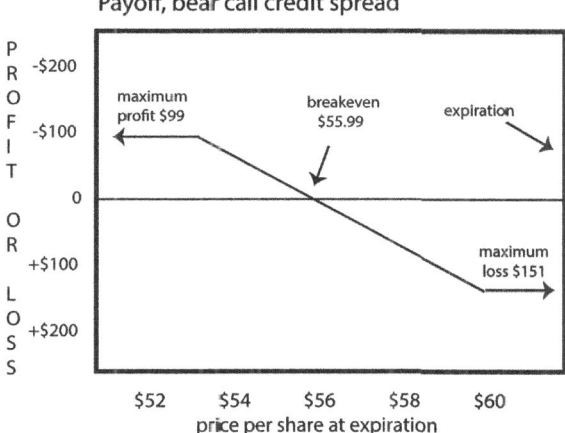

Fig. 7.4 Payoff, bear call credit spread—prepared by the author

Risk and Payoff Calculations: Condors and Butterflies

Moving beyond the spread with its two-position offsets, more complex spreading strategies mix limited risk with limited profit, regardless of the direction the underlying moves. This is accomplished through a range of condor and butterfly positions.

In these strategies, closing positions in legs often is discouraged because it easily sets up higher risks in remaining positions left open. One exception to this has been observed, however, when underlying price movement is substantial, setting up a legging out decision:

> … that is acceptable and may even be prudent. Since the spread consists of both a bull spread and a bear spread, it may often be the case that the stock experiences a relatively substantial move in one direction or the other … and that the bull spread portion or the bear spread portion could be closed out near their maximum profit potentials.[6]

This potential change in the underlying highlights the distinction between profit, breakeven or loss at expiration, versus real-world possibilities of early close for part (or all) of a condor or butterfly.

The following examples assume positions are held intact until expiration. In practice, portions of these combined positions become profitable and will be closed, leaving open the remaining positions. Ideally the closed positions cover the net cost of the entire position, so that additional favorable price movement will set up more profit.

Condor

The condor consists of four positions and four strikes, combining four separate calls (or puts) and the combination of long and short positions. Using calls, a condor is established by trading two ITM call: selling one and buying another at a lower strike; and at the same expiration, trading two OTM calls: selling one and buying another at a higher strike.

Maximum profit is achieved at the net difference between the two lowest strikes, minus the net debit paid:

$$S_s - S_l - (D - F) = P$$

S_s strike, lower short call
S_l strike, lower long call
D debit paid
F trading fees
P maximum profit

Breakeven occurs at two different points, one each for the upper strikes and lower strikes:
Upper breakeven: $S_u - (P - F) = B_u$
Lower breakeven: $S_l + (P - F) = B_l$

S_u Highest long strike
P Premium paid
F trading fees
B_u Upper breakeven
S_l Lowest long strike
B_l Lower breakeven

Maximum loss is limited to the net debit paid to open the position. It is realized when one of two underlying proximity events occurs: When the underlying price is equal to or lower than the lower long strike, or when it is equal to or higher than the higher long strike:

$$P + F = L$$

Table 7.3 Calls and puts, Macy's (M)—prepared by the author

Strike[a]	Calls		Puts	
	Bid	Ask	Bid	Ask
Dec 16:				
40	2.20	2.32	1.06	1.12
41	1.60	1.68	1.49	1.55
42	1.11	1.20	2.02	2.13
43	0.73	0.80	2.67	2.81

[a]M's price at this time was $41.36, at the close on November 11, 2016

P Premium paid
F Trading fees
L Maximum loss

For an example of a condor trade, Table 7.3 presents options values for Macy's (M).

A condor is constructed with the following positions:

Sell 41 call, bid 1.60, less trading fees = −$151
Buy 40 call, ask 2.32, plus trading fees = $241
Sell 42 call, bid 1.11, less trading fees = −$102
Buy 43 call, ask 0.80, plus trading fees = $89
Net debit = $77

The outcomes based on these positions:

Maximum profit: (41 − 40) − 0.77 = 0.23 ($23)
Breakeven (upper): 43 − 0.77 = 42.23 ($42.23)
Breakeven (lower): 40 + 0.77 = 40.77 ($40.77)
Maximum loss: $241 + $89 − $151 − $102 = $77

These outcomes are illustrated in Fig. 7.5.

Iron Condor

The *iron condor* acquires the same shape and appearance as the condor, but sets up as a net credit rather than as a debit. It combines a bull put spread

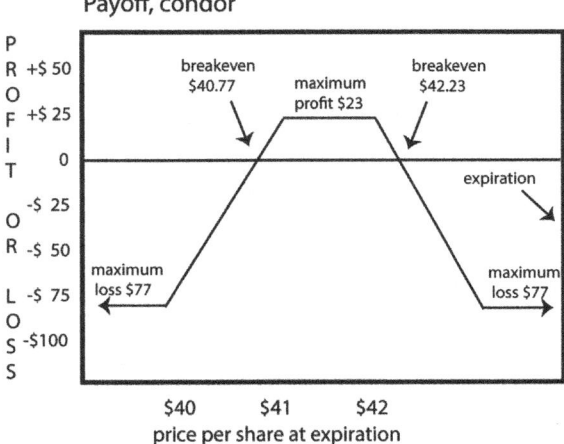

Fig. 7.5 Payoff, condor—prepared by the author

with a bear call spread. The four positions are a short OTM put with a long, lower-strike put; and a short OTM call with a higher-strike call.

Maximum profit is limited to the net credit received:

$$P - F = M$$

P Premium received
F Trading fees
L Maximum profit

In the iron condor, two breakeven prices occur, one above and one below:

$$S_c + P = B_u$$

$$S_p - P = B_l$$

S_c Strike, short call
S_p Strike, short put
P Net premium received
B_u Upper breakeven
B_l Lower breakeven

A loss occurs when the underlying price declines below the lower put strike, or equals or exceeds the higher long call strike. This loss is identical

on either side and is limited to the net strike distance between the options on either side, less the net credit received. Maximum loss:
$$S_u - B_u = M \text{ or } B_l - S_l = M$$

S_u strike, upper long call
S_l strike, lower long call
M maximum loss

For an example of the iron condor, the previously introduced Table 7.3 is used for options on Macy's (M). An iron condor is constructed with the following positions:

Sell 41 put, bid 1.49, less trading fees = −$140
Buy 40 put, ask 1.12, plus trading fees = $121
Sell 42 call, bid 1.11, less trading fees = −$102
Buy 43 call, ask 0.80, plus trading fees = $89
Net credit = $32

Outcomes for the iron condor in this example:

Maximum profit: ($140 + $102) − ($121 + $89) = $32
Upper breakeven: 43 + 0.32 = 42.32 ($42.32)
Lower breakeven: 41 − 0.32 = 40.68 ($40.68)
Maximum upper loss: 43 − 42.32 = 0.68 ($68)
Maximum lower loss: 40.68 − 40 = 0.68 ($68)

These outcomes are illustrated in Fig. 7.6.

Long Butterfly

The butterfly is very similar to the condor. It is a neutral strategy combining a bull spread with a bear spread. Unlike the condor's four strikes, the butterfly has only three.

The long call butterfly is a combination of one ITM position, two ATM, and one OTM. Maximum profit will be realized when the underlying remains ATM by expiration. The formula for maximum profit is:
$$S_s - S_t - (P - F) = M$$

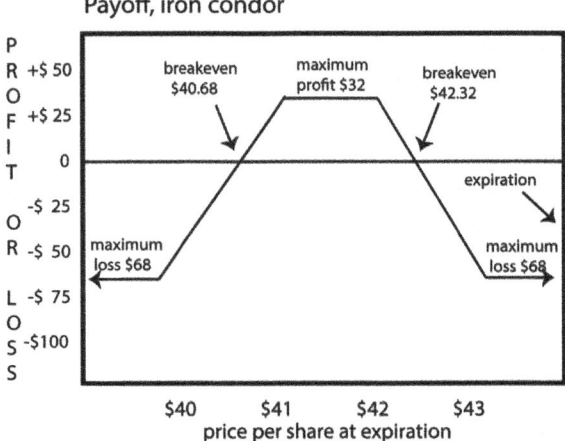

Fig. 7.6 Payoff, iron condor—prepared by the author

S_s strike, short call
S_l strike, lower long call
P premium paid
F trading fees
M maximum profit

Breakeven occurs at two points, one above and the other below:

$$S_u - (P - F) = B_u$$
$$S_l + (P + F) = B_l$$

S_u strike, higher long call
S_l strike, lower long call
P premium paid
F trading fees
B_u upper breakeven
B_l lower breakeven

Maximum loss is equal to the initial net debit paid for the position. This occurs in one of two situations:

$$U \geq S_h$$
$$U \leq S_l$$

- U underlying
- S_h strike, higher long call
- S_l strike, lower long call

The maximum loss is calculated as:

$$P - F = M$$

- P premium paid
- F trading fees
- M maximum loss

To create an example of the long butterfly, Table 7.4 summarizes three strikes:

A long butterfly is created using calls with the following:

Buy one 62.50 call, ask 3.50, plus trading fees = $359
Sell two 65 calls, bid 1.78 each (total $356), less trading fees[1] = −$346
Buy one 67.50 call, ask 0.86, plus trading fees = $95
Net debit = $108

The possible outcomes for this butterfly are:

Maximum profit: (65 − 62.50) − 1.08 = 1.42 ($142)
Upper breakeven: 67.50 − 1.08 = 66.42 ($66.42)
Lower breakeven: 62.50 + 1.08 = 63.58 ($63.58)
Maximum loss: $359 + $95 − $346 = $108

These outcomes are illustrated on Fig. 7.7.

Table 7.4 Calls and puts, Occidental Petroleum (OXY)—prepared by the author

Strike[a]	Calls		Puts	
	Bid	Ask	Bid	
Dec 16:				
62.50	3.30	3.50	1.35	1.43
65	1.78	1.87	2.38	2.53
67.50	0.79	0.86	3.95	4.25

[a]OXY's price at this time was $64.95, at the close on November 11, 2016

---

[1]For two options, the trading fee is rounded to $10, versus rounded $9 for a single option.

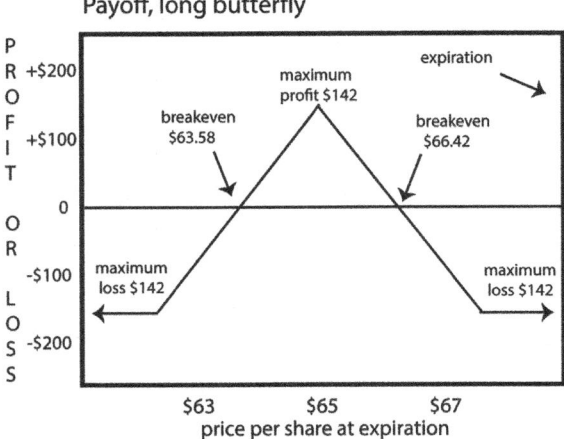

Fig. 7.7 Payoff, long butterfly—prepared by the author

Short Butterfly

The short butterfly reverses the long and short for each of the three strikes and can be built with either calls or puts. Constructed with calls, a short butterfly consists of one ITM short call, two ATM long calls, and one OTM short call.

Maximum profit is accomplished when one of two situations arises:

$$U \geq S_u \quad or \quad U \leq S_l$$

U underlying
S_u strike, upper short call
S_l strike, lower short call

The dollar amount of maximum profit is limited to the credit received, minus trading fees:

$$P - F = M$$

P premium received
F trading fees
M maximum profit

Breakeven occurs at two separate positions, one above and the other below:

$$S_u - (P - F) = B_u$$
$$S_l + (P - F) = B_l$$

S_u strike, highest short call
S_l strike, lowest short call
P premium received
F trading fees
B_u upper breakeven
B_l lower breakeven

A loss is limited to a middle range of underlying price remaining unchanged by expiration. This level occurs when:

$$U = S$$

U underlying
S strike, long calls

The formula for maximum loss is:

$$S - S_l - (P - F) = M$$

S strike of long calls
S_l strike, lower strike short call
P premium received
F trading fees
M maximum loss

The short butterfly can be created using options shown in the previously introduced Table 7.4.

A summary:

Sell one 62.50 call, bid 3.30, less trading fees = −$321
Buy two 65 calls, ask 1.87 (3.74), plus trading fees[2] = $384
Sell one 67.50 call, bid 0.79, less trading fees = −$70
Net credit = $7

The outcomes for this short butterfly are:

Maximum profit: $321 − $384 + $70 = $7
Upper breakeven: 67.50 − 0.07 = 67.43 ($67.43)
Lower breakeven: 62.50 + 0.07 = 62.57 ($62.57)
Maximum loss: 65 − 62.50 − 0.07 = 2.43 ($243)

---

[2]For two options, the trading fee is rounded to $10, versus rounded $9 for a single option.

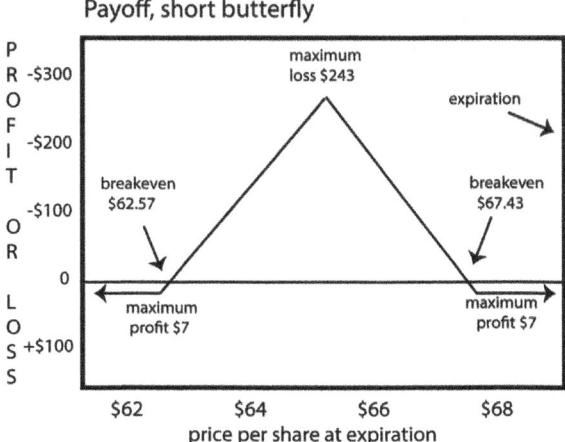

Fig. 7.8 Payoff, short butterfly—prepared by the author

Based on the comparative likely outcomes (maximum profit of $7 versus maximum loss of $243), the short butterfly in this example does not represent an acceptable risk. However, by selecting different strikes, potential profits and losses could be altered. In addition, the outcomes are modeled on the basis of keeping all positions open until the last trading day. This is necessary for comparative purposes; however, in practice, short sides are likely to be closed to take profits due to time decay and long positions are likely to be closed if intrinsic value increases during the open period.

Because the short positions reside in the short butterfly at the highest and lowest strikes, traders need to carefully monitor underlying price behavior. The short positions may be closed (on one side or the other) if it appears likely that the short options are going to move in the money. The timing of closure may be designed to exploit time decay. As expiration approaches, two key dates should be kept in mind for timing a buy to close order, the Friday before expiration and the day before expiration: "During the weekend that precedes expiration, an option contract loses 32.8% of its remaining time. Thursday evening before the final trading day, 31.3% of the remaining time disappears."[7]

The sample outcomes are summarized in Fig. 7.8.

Iron Butterfly

The iron butterfly is creating by combining call and put spreads, with each side consisting of OTM and ITM strikes: equal numbers of long OTM and short ITM puts, and of short ITM calls and long OTM calls.

Maximum profit is possible when the underlying price is equal to the strike of the short call and put midway in the strike positions. The formula is equal to the net credit received:

$$P - F = M$$

P premium received
F trading fees
M maximum profit

Breakeven occurs at one of two price levels, one above and one below:

$$S_c - (P - F) = B_u$$

$$S_p - (P - F) = B_l$$

S_c strike of short call
S_p strike of short put
P premium received
F trading fees
B_u upper breakeven
B_l lower breakeven

Loss is also limited, and occurs in one of two situations:

$$U \geq S_c$$

$$U \leq S_p$$

U underlying
S_c strike, long call
S_p strike long put

The formula for maximum loss is:

$$S_l - S_s - (P - F) = M$$

S_l strike, long call
S_s strike, short call
P premium received
F trading costs
M maximum loss

An iron butterfly can be constructed based on previously introduced Table 7.4:

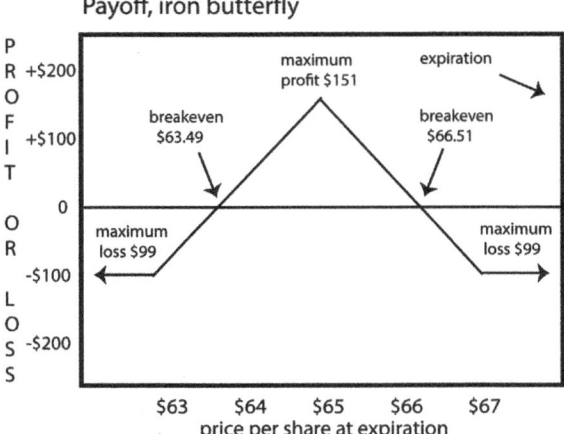

Fig. 7.9 Payoff, iron butterfly—prepared by the author

Buy one 62.50 put, ask 1.43, plus trading fees = $152
Sell one 65 put, bid 2.38, less trading fees = −$229
Sell one 65 call, bid 1,78, less trading fees = −$169
Buy one 67.50 call, ask 0.86, plus trading fees = $95
Net credit = $151

The outcomes based on this set of trades:

Maximum profit: $229 + $169 − $152 − $95 = $151
Upper breakeven: 65 + 1.51 = 66.51 ($66.51)
Lower breakeven: 65 − 1.51 = 63.49 ($63.49)
Maximum loss: 67.50 − 65 − 1.51 = 0.99 ($99)

These outcomes are illustrated in Fig. 7.9.

Risk and Payoff Calculations: Synthetics

A synthetic stock position is so named because the net value of the options mirrors price activity in the underlying. Thus, a synthetic long stock position is most advantageous when the underlying price rises. The intrinsic value of the long call will rise point for point with the rise in underlying price (and the short put will expire worthless). A synthetic short stock mirrors stock price movement as well. If the underlying price declines, the long put gains one point for each point in decline of the underlying, and the short call will expire worthless. In both strategies, a move in the wrong direction (underlying declining with a synthetic long stock, or rising with a synthetic short stock), the mirroring is represented in accumulating net losses.

Although the synthetic long and short positions both mirror price movement in the underlying, the risk levels are significantly different. The synthetic long stock involves a long call and a short put; and the short put's market risk is identical to market risk of a covered call. This makes the synthetic long stock a conservative strategy, since the uncovered put is easily managed and either closed or rolled to avoid exercise. However, the synthetic short stock involves an uncovered call, making it a much higher-risk strategy. This high risk is eliminated, however, if the synthetic short stock is opened along with ownership of 100 shares of the underlying, converting the uncovered call to a more conservative covered call.

Synthetic Long Stock

The synthetic long stock strategy is a bullish combination of a long call and a short put, at the same strike and expiration. Ideally, both should be at the money. Maximum profit is unlimited, since the long call has the potential to grow in value indefinitely if and when the underlying value rises. The formula for profit is calculated in one of two ways, depending on whether the net options set up as a debit or a credit:

$$U - S_c - (P + F) = M \:(with\:net\:debit)$$

$$U - S_c + (P - F) = M \:(with\:net\:credit)$$

U underlying price per share
S_c strike, long call
P premium paid
F trading fees
M maximum profit

Breakeven is also calculated in one of two ways, depending on the net debit or credit of the synthetic options:

$$S_c + (P + F) = B \:(with\:net\:debit)$$

$$S_c - (P - F) = B \:(with\:net\:credit)$$

S_c strike, long call
P premium paid
F trading fees
B breakeven

The synthetic long stock position has unlimited loss, which is realized if and when the underlying declines. Although this is called "unlimited," in fact it is limited to the maximum level to which the underlying may decline, which is zero. The formula for loss is calculated differently when a debit is incurred, versus when a credit is received:

$$S_p - U + (P + F) = M \text{ (with net debit)}$$

$$S_p - U - (P - F) = M \text{ (with net credit)}$$

S_p strike, short put
U underlying price
P premium paid
F trading fees
M net loss

This position can be set up with options on Kellogg (K) as shown on Table 7.5.

Using ATM positions:

Buy 72.50 call, ask 2.00, plus trading fees = $209
Sell 72.50 put, bid 2.25, less trading fees = −$216
Net credit—$7

Based on this combination of positions, the outcomes, based on assumed potential underlying moves of three points above or below the strikes, are:

Table 7.5 Calls and puts, Kellogg (K)—prepared by the author

Strike[a]	Calls		Puts	
	Bid	Ask	Bid	Ask
Dec 16:				
70	3.20	3.60	1.15	1.35
72.50	1.85	2.00	2.25	2.50
75	0.95	1.05	3.80	4.20

[a]K's price at this time was $72.50, at the close on November 11, 2016

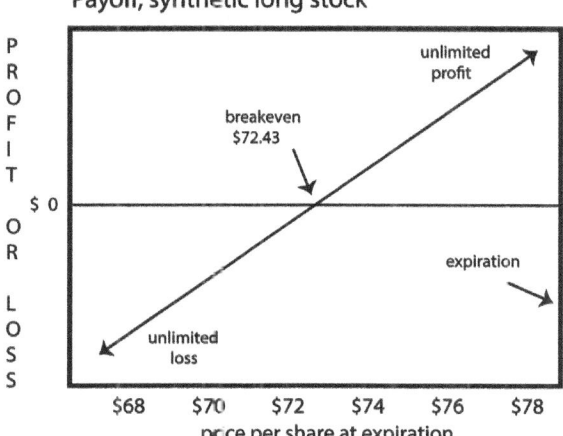

Fig. 7.10 Payoff, synthetic long stock—prepared by the author

Profit[3]: $75.50 - 72.50 + 0.07 = 3.07$ ($3.07)
Breakeven: $72.50 - 0.07 = 72.43$ ($72.43)
Loss[4]: $72.50 - 69.50 - 0.07 = 2.93$ ($293)

This range of outcomes is illustrated in Fig. 7.10.

Synthetic Short Stock

The opposite of synthetic long stock is the synthetic short stock. This is a bearish combination of a long put and a short call. The potential loss is assumed to be unlimited as long as the short call is uncovered. However, if the synthetic short stock is entered along with ownership of 100 shares, the potential loss is limited to the net difference between the call's strike and the original cost of shares, increased for a net debit or decreased for a net credit in the option position. The ideal situation is for the strike to be exactly at the money.

Profit is unlimited. However, as a practical matter, the underlying price can decline only to a finite level (zero), which limits the "best-case" outcome for the bearish position. The profit is achieved when the underlying price is lower than the long put's strike, plus a net credit, or minus a net debit for opening the position:

---

[3]This outcome is based on assumed three-point move in the underlying above the strikes.
[4]This outcome is based on assumed three-point move in the underlying below the strikes.

$$S_p - U - (P + F) = M \text{ (with net debit)}$$

$$S_p - U + (P - F) = M \text{ (with net credit)}$$

- S_p strike, long put
- U underlying price per share
- P premium paid
- F trading fees
- M maximum profit

Breakeven is calculated based on whether the short synthetic creates a net credit or debit:

$$S_p - (P - F) = B \text{(with net credit)}$$

$$S_p + (P + F) = B \text{(with net credit)}$$

- S_p strike, long put
- P premium paid
- F trading fees
- B breakeven

A net loss is realized when the underlying price rises. This results in exercise of the short call unless that position is closed or rolled forward. The risk is further mitigated if you also own 100 shares of the underlying for each short option, converting the call to a covered position. A maximum loss occurs when the underlying price is greater than the short call's strike, adjusted for the net premium:

$$U - S_c + (P + F) = L \text{ (with net debit)}$$

$$U - S_c - (P - F) = L \text{ (with net credit)}$$

- U underlying
- S_c strike, short call
- P premium
- F trading fees
- L loss

A synthetic short stock can be set up with the previous Table 7.5, with the following positions:

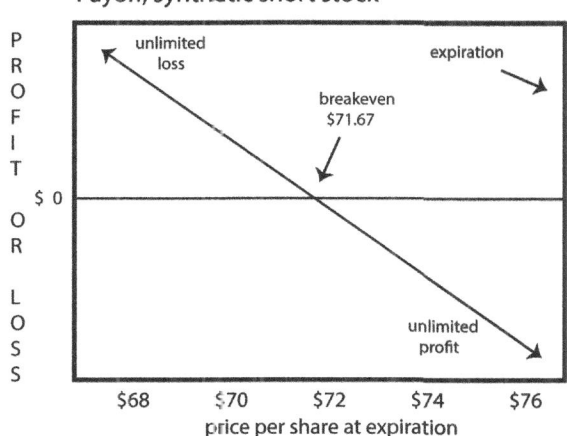

Fig. 7.11 Payoff, synthetic short stock—prepared by the author

Buy 72.50 put, ask 2.50, plus trading fees = $259
Sell 72.50 call, bid 1.85, less trading fees = −$176
Net debit = $83

Outcomes for this position are:

Profit[5]: 72.50 − 69.50 − 0.83 = 2.17 ($217)
Breakeven: 72.50 − 0.83 = 71.67 ($71.67)
Loss[6]: 75.50 − 72.50 + 0.83 = 3.83 ($383)

These outcomes are illustrated in Fig. 7.11.

Risk and Payoff Calculations: Horizontal Spreads

The horizontal spread (calendar spread) offsets long and short positions with the same strike and different expirations. The rationale is based on two assumptions. Opening a shorter-term short position translates to a more rapid time decay, versus a longer-term long position at the same strike. It should be possible to profit from time decay by opening this position.

---

[5] This outcome is based on assumed three-point move in the underlying below the strikes.
[6] This outcome is based on assumed three-point move in the underlying above the strikes.

The second assumption is that the shorter-term short position is covered by the same-strike long options. As a result, in the even the short side is exercised, the long positions can be used to satisfy the assignment.

Using calls for spreads is a bullish strategy, since the longer-term long calls would profit if and when the underlying rises above the strike. Puts represent a bearish version of the calendar spread, assuming a trader expects the underlying to fall below the long strikes before they expire.

Bull Calendar Spreads

The bull calendar spread consists of calls OTM by a minimal number of strikes, preferably only one or two. This maximizes the option price reaction to price movement in the underlying. In the ideal configuration, the short-term short positions pay for most of the cost of the long-term options. This translates to a requirement for very little price movement to meet and exceed the breakeven point.

Profit potential is unlimited, based on the best outcome: the short-term short positions remaining at or out of the money until expiration, followed by movement in the money before the long positions expire. Profit is realized when the underlying exceeds the long strike. The formula for maximum profit is:

$$U - S - (P + F) = M$$

- U underlying price by expiration
- S strike
- P premium
- F trading fees
- M maximum profit

Breakeven relies on the timing between entry and expiration for both sides of the trade. Assuming the short-term options expire worthless, breakeven equals the amount the long call moves in the money:

$$S + (P + F) = B$$

- S strike of the long call
- P debit paid for the long call
- F trading fees
- B breakeven, long call

Loss is limited to the amount of the original debit paid for the spread:

$$P + F = M$$

P debit paid for the position
F trading fees
M maximum loss

An example of the bull calendar spread using calls is based on the premium values summarized in Table 7.6.

A spread is established with the following positions:

Sell 12/16 call 29 strike, bid 0.34, less trading fees = −$25
Buy 12/23 call 29 strike, ask 0.47, plus trading fees = $56
Net debit = $31

Calculating outcomes based on this trade:

Profit[7]: 30 − 29 − 0.31 = 0.69 ($69)
Breakeven: 29 + 0.31 = 29.31 ($29.31)
Loss: $56 − $25 = $31

This range of outcomes is illustrated in Fig. 7.12.

Employing puts for a calendar spread sets up an opposite directional move and is entered in the belief that the long-term long puts will appreciate based on a decline in the underlying below the strike.

The calendar spread may be varied in several ways, including setting up ratios with weighted shorts over lower numbers of long options, or in the form of backspreads. Traders may not be aware, however, that the ratio is not merely a way to adjust the net cost of a spread. It also increases risks. When a trader first analyzes a trade with limited loss exposure in mind, the introduction of a ratio changes that dynamic:

> It is critical to understand that while ratio spreads are obviously called "spreads," the options are really only partially spread. Because the quantity is different for the number of options bought and sold, the excess on either side are either left long or uncovered (naked). This is critically important because while most spreads have a limited loss potential, any ratio spread that includes uncovered options will have an unlimited or significant loss potential.[8]

---

[7]Profit assumes the underlying moving one point in the money.

Table 7.6 Calls and puts, MGM Resorts (MGM)—prepared by the author

Strike[a]	Calls		Puts	
	Bid	Ask	Bid	Ask
Dec 16:				
27	1.23	1.27	0.74	0.76
27.50	1.05	1.10	1.01	1.08
28	0.71	0.74	1.20	1.23
29	0.34	0.38	1.79	1.88
30	0.17	0.18	2.56	2.72
Dec 23:				
27	1.40	1.46	0.85	0.94
27.50	1.12	1.18	1.07	1.17
28	0.80	0.84	1.25	1.34
29	0.43	0.47	1.88	2.00
30	0.21	0.26	2.61	2.77

[a]MGM's price at this time was $27.50, at the close on November 11, 2016

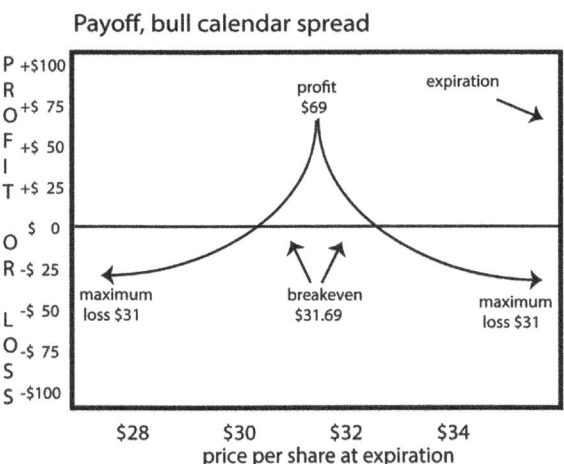

Fig. 7.12 Payoff, bull calendar spread—prepared by the author

The possible combinations and variations make the realm of horizontal spreads diverse with a range of risks and risk mitigation greater than any other structure. For example, the calendar spread can be set up to hedge equity positions over an extended period of time, with the installment calendar spread strategy.

Installment Calendar Spreads

The "installment" version of the calendar spread is so called because it sets up a two-part trade, one many months away and the other with a very short term until expiration. The longer-term long option is paid for in a series of installments consisting of short calls or puts.

The uncovered put is by definition a conservative trade, as its market risk is identical to the market risk of a covered call. A short call may be covered or uncovered. However, ownership of a long-term long call provides an alternative form of cover, mitigating the short call market risk. Another way to manage risk for these short positions is through selection of strikes with buffer zones, making exercise less remote. With the buffer zone as well as very rapid time decay, the likelihood of a successful installment series is quite high. The test of overall risk is determining the number of times a short option has to be traded in order to pay for the longer-term long option.

There are two versions of the installment spread, involving a LEAPS call or put. In addition to these basic forms of the installment strategy, the positions can be modified with the use of straddles or strangles, diagonal positions, and other possible modifications to change risk levels or to accelerate the time required to pay for the long-term option. The strategy is expanded by opening multiple long positions and then varying the offsetting short side. For example, opening two long positions may be offset by opening both a short call and a short put (or two positions on one side or the other). This adds to the flexibility of the installment calendar spread.

The "contingent purchase" installment is a calendar spread combining the purchase of a LEAPS call with a series of short-term short calls or puts (or both). This strategy makes sense for a trader who wants to purchase shares of the underlying but cannot afford to pay for them today; or who is uncertain about the short-term price direction. In either case, freezing the contingent purchase price is advantageous. It sets up the ability to buy shares at a later date (or not), and to pay for the long call over a series of short option sales.

For example, Chipotle (CMG) closed on November 11, 2016 at $397.08 per share, as summarized in Table 7.7.

You are interested in buying shares in the future, but rather than doing so today, you enter a contingent purchase strategy with the following positions:

Buy 400 strike call expiring June 16, 2017, ask 43.80, plus trading fees = $4,389

Sell December 16, 2016 short options, either:
4.20 call, ask 6.30, less trading fees = $621
3.80 put, ask 8.40, less trading fees = $831

This selection sets up a 20-point buffer between the stock price of $397.08 and the strike of each option. The strategy depends on time decay between entry date and expiration. However, once time decay takes the option down so that most of the time decay has disappeared, the short option can be bought to close and replaced with a later-expiring contract.

Is this a viable strategy? To determine this, divide the cost of the LEAPS call by the net credit of the short-term option:

$$P_l \div P_s = N$$

P_l premium, long option
P_s premium, short option
N number of turns required

Assuming the current premium is typical of what can be opened throughout the 219-day period between the entry date and LEAPS expiration, this calculation reveals the number of similar trades required to pay for the long option. In the case of the short call: $4389 ÷ $621 = 7.07 *times*.

A trade with similar premium has to be executed 7.07 times to pay for the LEAPS call. The next step is to determine how frequently such a trade must be entered:

$$D \div T = F$$

D days until LEAPS expiration
T number of times required
F frequency required

The long call with similar pricing would have to be executed at least once every 31 days, or on average once per month:

$$219 \div 7.07 = 31$$

The same set of calculations can also be applied to the short put: $4389 ÷ $831 = 5.28 *times* and 219 ÷ 5.28 = 41

Executing a similar short put strategy once every 41 days will pay for the LEAPS call, or on average once every six weeks.

At the end of the term, several possible actions result. If the underlying price is lower than the call's strike, it expires worthless. Because this long call was paid for with a series of short calls and/or puts, nothing was lost; and the trade fixed the price of 100 shares in the event the price had moved higher.

If the price has moved higher, you can sell the call and take profits; or you can exercise the call and purchase 100 shares at the fixed strike, which is lower than current market value.

This strategy allows you to fix the future price of shares for no net cost. It avoids lost opportunity. If the stock price moves upward throughout the period, having taken no action translates to a lost opportunity to buy shares. The installment is a rational alternative because it results in a zero-basis long call that otherwise would have been very expensive.

The second installment calendar spread is the offset between a long put and a series of short calls and/or puts. This is a risk hedge, appropriate for investors who own shares and who want to eliminate market risk. A LEAPS put is purchased to freeze risk at the strike into the future. Based on the same options introduced in Table 7.7, a seven-month risk freeze is accomplished with the installment risk hedge. For example, this strategy is set up with the following positions:

Table 7.7 Calls and puts, Chipotle (CMG)—prepared by the author

Strike[a]	Calls		Puts	
	Bid	Ask	Bid	Ask
12-16-16:				
360	40.90	41.60	4.00	4.40
370	32.70	33.30	5.90	6.20
380	25.40	25.90	8.40	8.70
390	18.90	19.40	11.90	12.30
400	13.60	14.00	16.50	16.90
410	9.30	9.70	22.20	22.70
420	6.30	6.70	29.10	29.60
430	4.10	4.50	36.90	37.60
440	2.75	3.10	44.40	47.30
6-16-17:				
395	43.20	46.60	39.10	42.30
400	40.60	43.80	42.10	44.90
405	38.10	41.80	44.60	47.40

[a]CMG's price at this time was $397.08, at the close on November 11, 2016

Buy June 16, 2017 400 put, 44.90, plus trading fees = $4499
Sell December 16 short options, either:
Call, 4.20 bid 6.30, less trading fees = $621
Put, 380 bid 8.40, less trading fees = $831

The same calculations are performed in this case. For the call:

$4499 \div $621 = 7.24 *times* and the number of days : 219 \div 7.24 = 30

A similar trade is required once every 30 days.

For the short put, the two calculations are:
$4499 \div $831 = 5.41 *times* and 219 \div 5.41 = 40

Similar trades have to be made at least once every 40 days when using puts.

The selection of short calls or short puts can be based on proximity between current price of the underlying, and the trading range. Using a buffer (in the previous examples, a 20-point buffer was applied), a short put is most likely to be selected when the underlying price is at the low swing within the trading range; and a short call would be more appropriate when the underlying price is close to the top of the swing.

To maximum the advantageous timing of short options, a trade should be entered only when a reversal signal is located and confirmed. Reliance on the Bollinger Bands, candlestick signals, volume, and momentum signals improves timing for these trades.

Risk and Payoff Calculations: Diagonal Spreads

In a diagonal spread, the attribute of vertical spread (same expiration) and calendar spread (same strike) are combined, to create a spread with different expiration and strike. This can be set up using either calls or puts.

The call diagonal consists of an equal number of longer-term long ITM calls and shorter-term OTM short calls with higher strikes. The trader's expectation for this trade is that the underlying price will not move in the money during the life of the short contracts; and will then move higher to exceed the long call strikes before expiration. When this occurs, the short calls expire worthless and the later-expiring long calls become profitable.

The strategy combines a long calendar call spread with a short call spread. The two-phase strategy depends on initial time decay followed by a bullish move in the underlying. Profit is limited and consists of the premium received for selling the earlier calls (assuming worthless expiration or profitable buy-to-close), plus profits received for closing appreciated long calls. Because there are variables on both sides, the maximum profit relies on price behavior before expiration of both sides.

Breakeven occurs at two points. First is when the ITM value of short calls is equal to the premium received. Second is when the opposite-moving KITM value of long calls is equal to the net debit paid for that position.

Loss is also limited. If the position is set up as a net credit, it is equal to the differences between the short and long strikes, minus the credit. If a debit, the maximum loss is the difference between the two strikes, plus the net debit paid for the position. The same argument for met loss applies in the diagonal: Variables and a decision to close or await expiration affects the realized loss.

The put diagonal spread is the opposite. It is based on a premise that the shorter-term puts will expire worthless of can be closed at a profit (expecting the underlying to remain unchanged) and that the underlying price will then move downward to convert longer-term long puts into profitable status. The call diagonal is bullish and the bear version is bearish based on these expectations of underlying price behavior ∗∗∗.

Spreads offer exceptional variation in construction and duration. As a result, the risk level also varies considerably based on strike distance, time to expiration, and exposure of short positions. This is where the calculations of profit, breakeven and loss are most valuable. These allow you to quantify levels of risk as well as opportunity in the many forms of spread. Beyond those explained here, many further expansions of spread positions are also possible.

Chapter Summary

- vertical spread construction affects overall risk of either bull or bear positions
- risk and profit potential are dissimilar for debit and credit spreads
- a proper analysis of the condor reveals the rationale for this position
- differences between condors and butterflies are subtle but may be significant
- long and short synthetics are similar in construction, but risks are very different
- horizontal and diagonal spreadsdiagonal spreads may not be advantageous in some situations

Notes

1. Ma, Christopher K. & Rao, Ramesh P. (1988). Information asymmetry and options trading. *The Financial Review, 23*(1), 39.
2. Vejendla, Ajitha & Enke, David (2013). Performance evaluation of neural networks and garch models for forecasting volatility and option strike prices in a bull call spread strategy. *Journal of Economic Policy and Research, 8*(2), 1–19.
3. Enke, David & Amornwattana, Sunisa (2008). A hybrid derivative trading system based on volatility and return forecasting. *The Engineering Economist, 53*(3), 259–292.
4. Han, Ki C, PhD., CFP & Heinemann, Alexis (2008). A bull call spread as a strategy for small investors? *Journal of Personal Finance, 6*(2), 108–127.
5. McMillan, Lawrence G. (2002). *Options as a strategic investment.* 4th ed. New York NY: New York Institute of Finance. p. 443.
6. Ibid., p. 207.
7. Augen, Jeff (2009). *Trading options at expiration.* Upper Saddle River NJ: FT Press. p. 41.
8. Frederick, Randy (2007, March). Trading with ratio spreads. *Futures, 36*, 48–51.

8

Strategic Payoff: Straddles

Chapter Objectives.

- Analyze long and short straddle risks based on varying levels of volatility
- Evaluate how straddles combine bullish and bearish sides in single trades
- Study the risk elements of the covered straddle
- Compare straddle risks to strangle risks
- Articulate similarities between strips and strikes, both long and short
- Determine how strategic selection and timing of strikes reduces short option risks.

Among the possible options combinations, straddles are the highest-risk. A long straddle demand substantial price movement to cover initial cost as well as lead to profits; and short straddles end up with one side in the money by expiration.

Some traders select straddles and other combinations in a belief that profit potential is high. Other traders rely on current underlying volatility to time straddle trades. An alternative to this selection is to employ both methods:

> The combination of using both volatility and return forecasting provides more useful information than using either the volatility or return forecasting in isolation, leading to higher trading performance when investing options *(sic)* … options should have appropriate characteristics that match the applied trading strategy and investment time frame in order to increase the chance of gaining a positive outcome.[1]

---

[1] Upper profit calculation assumes underlying price increase of eight points, or $113 per share.

Long straddles often are timed and entered immediately before earnings reports if and when the cost of the straddle is considered relatively low. The rationale is that a big price move in the underlying is likely (based on past earnings surprises), but the direction of the move is not known. One side or the other in a straddle might become profitable enough to surpass the cost of the straddle. This is highly speculative. However, the rationale for a long straddle is difficult to identify other than when big changes are expected, due to earnings, announcements, or expected marketwide price changes.

Short straddles are most likely to become profitable when entered with exceptionally short time remaining until expiration. As time decay strategies, the trader hopes for two concurrent situations: Little or no movement in the underlying, combined with rapid time decay. Thus, short straddles for expirations of two weeks or less reduce exposure time. However, since one side is invariably in the money by expiration, closing at a profit is a logical expectation. One side—the one out of the money—will certainly lose time value and become profitable. The other side of the straddle requires time decay to more than offset increasing intrinsic value in order to become profitable.

The often marginal profitability of either long or short straddles reflects the consequence of focusing on ATM options at the time of entry. It makes sense given the potential underlying movement in *either* direction. However, this ATM status may be compared to the more favorably priced strangle, in which both sides normally are OTM, thus lower cost (for long strangles) or lower risk (for short strangles), and also for the European trade known as a double, defined as:

> A commodity option traded in Europe that gives its owner the right either to call (buy) or to put (sell) the underlying asset but not both. When one side of the option is exercised, the opposite side is automatically terminated.[2]

The overall cost and risk comparison is the means for judging the payoff calculations for straddles, strangles, and related combinations and also with the moneyness of positions in mind:

> Straddles, strangles, and doubles are combinations in the narrow sense of the term in that the trader buys or sells two options. All the others are spreads, in that the trader buys one or more options and sells one or more, so that the prices offset to some extent. Consequently, we expect straddles, strangles, and doubles to have much higher net prices than the spreads, and they do.

---

[2]Lower profit calculation assumes underlying price decrease of 15 points, or $90 per share.

The net price also reflects whether these trades normally involve in- or out-of-the-money contracts and the times to expiration. For instance, the net price for straddles is more than double that for strangles because in a straddle one of the options is always in the money while with strangles both legs are usually out of the money, and because straddles generally have longer expirations.[3]

The question of cost and risk is further complicated by how traders may control risk exposure through ownership of the underlying security. Thus, the obvious high risk of a short straddle or strangle is vastly changed by owning shares, in which case the exposed call is covered, converting from a high-risk to a low-risk posture. A motive for owning shares might initially be to reduce option-related risk; however, this also sets up a hedging mechanism for the equity position. Hedging is more popular with options than the historical speculation associated with trading, notably in the utilization of short combinations. In this regard, hedging and its separate risks often are managed and even offset through options trades. One study highlighted this by observing that:

> the correlation between the accumulated hedging errors for different options can be quite high, so that the risk of arbitrage due to hedging errors can be substantially reduced by optimally combining options into portfolios.[4]

With these observations concerning risk and cost, straddle payoff calculations should be recognized as a consistent analysis for outcomes in the event that positions are held open until the last trading day. In practice, early closing to take profits, rolling short positions forward to avoid exercise, and acceptable of exercise as a result, are all potential alternative outcomes.

Risk and Payoff Calculations: Straddles

The straddle is vastly different in attributes than the spread. Whereas a spread can be structured to create profits based on hedging, a straddle's success or failure relies more heavily on volatility of the underlying asset. For traders focused on implied volatility (as an *estimate* of future volatility) the risk of accurately timing a straddle is considerably higher than for those focused on the underlying asset, and its *known* historical volatility. This is due to the less reliable nature of both implied volatility, versus historical volatility: "Implied

---

[3]Upper loss calculation assumes underlying price increase of seven points, or $112 per share.
[4]Lower loss calculation assumes underlying price decrease of 8 points, or $97 per share.

standard deviations are biased forecasts of future volatility, and were found to be worse estimators than historical based volatility estimators."[5]

The nature of each form of volatility further points out how each is influenced by news and events. For traders in straddles, volatility is of immediate concern and interest, so the disparity between the two volatility measurements demonstrates why the historical alternative is a more accurate risk measurement:

> In the stock options market, earnings news, the potential success or failure of new products, or (most dramatically) the possibility of a takeover, can all cause increases in implied volatility, regardless of the historical volatility of the stock.[6]

Long Straddles

Assuming that weight is given to historical volatility, a long straddle can be treated as a special form of options trading:

> The returns of an option straddle depend on the volatility of the underlying asset. Hence buying an option is similar to "investing" in volatility.[7]

However, the significance of volatility has an opposite effect between long or short positions. Maximum long straddle profits are likely to be realized at times of high volatility (versus low volatility for short straddles). A related benefit to volatility-based trading is discovered in efficiency of option pricing: "Volatility trades ... tend to equalize expected and implied volatility, helping to ensure that derivative securities are correctly priced."[8]

Unlike clearly bullish or bearish trades, the long straddle combines both. The long call is bullish and the long put is bearish. This makes the straddle a directional hedge; one side or the other will gain intrinsic value and the trader hopes that movement is enough to cover the straddle's overall cost. Ideally, the long straddle is set up with ATM options, consisting of an equal number of calls and puts.

---

[5]The calculation of net loss assumes a decline in the underlying price to $92 per share and the put's last trading day intrinsic value of 10 points, or $1,000. It further assumes a loss on both the put and the stock, based on closing both positions. This presents a problem in analysis, since the worst case is not likely to occur. In practice, the stock does not have to be sold, and the put may be rolled forward to avoid exercise.
[6]The example for upper profit is based on assumed increase of 20 points in the underlying.
[7]The example for lower profit is based on assumed decrease of 25 points in the underlying.
[8]The example for upper loss is based on assumed increase of 20 points in the underlying.

Given the need for substantial price movement in the long straddle to accomplish breakeven and profitability, the timing of this trade has to be specific. Beyond the desired volatility effect making long straddles viable as a form of speculation, timing for earnings is also a likely trigger for timing. A company with a record of large earnings surprises may also exhibit past price gaps, often exaggerated; and then a correction in the opposite direction. A trader who believes this pattern is likely to be repeated in the current quarter may enter a long straddle. Believing a large surprise is likely, but not knowing the direction (positive or negative), the long straddle offers the potential for profits in either direction.

The attraction of the long straddle is that profits are unlimited, whereas maximum loss is limited. Maximum profit is accomplished when the underlying price is either higher than the long call strike, or lower than the long put strike, *and* when the level exceeds net cost of the position in either direction. On the bearish side, the potential profit is described as "unlimited," but in reality, there is a limit. An underlying price can decline only to zero at the extreme, and that poses a limit on the maximum downside potential of the long straddle.

The formula for maximum profit is:

$$U - S - (P + F) = P_u$$
$$S - U - (P + F) = P_l$$

U underlying price
S strike
P premium paid
F trading fees
P_u upper profit (when $U > S - P + F$)
P_l lower profit (when $S < U - P + F$)

Breakeven also has both an upper and a lower level:

$$S + P + F = B_u$$
$$S - P + F = B_l$$

S strike
P premium paid
F trading fees
B_u upper breakeven
B_l lower breakeven

Table 8.1 Calls and puts, Anheuser-Busch (BUD)

Strike [a]	Calls bid ask		Puts bid ask	
1/20/17				
95	10.60	11.20	0.45	0.65
100	6.40	6.90	1.20	1.30
105	3.20	3.40	2.90	3.10
110	1.25	1.40	5.80	6.20
115	0.45	0.60	9.90	10.40

[a]BUD's price at this time was $105.05, at the close on December 13, 2016

Maximum loss occurs when the underlying price is at the strike as of expiration. Both sides expire worthless in this situation and the loss equals the premium paid:

$$P + F = M$$

P premium paid
F trading fees
M maximum loss

For example, a long straddle can be created using the positions shown on Table 8.1.

Set up a long straddle with the positions closest to the money:

Buy 105 call, ask 3.40, plus trading fees = $349
Buy 105 put, ask 3.10, plus trading fees = $319
Total debit = $668

To compute possible outcomes:

Upper profit: 113–105–6.68 = $132[1]
Lower profit: 105–90–6.68 = $832[2]
Upper breakeven: 105 + 6.68 = $121.68
Lower breakeven: 105–6.68 = $98.32
Maximum loss: 6.50 +0.18 = 6.68 ($668)

This range of outcomes is also illustrated in Fig. 8.1.

The requirement for underlying price movement defines a long straddle as dependent on high volatility. In comparison, the short straddle is the opposite. It offers limited profits, and only if and when the underlying price does not move beyond a range tied to the ATM strike.

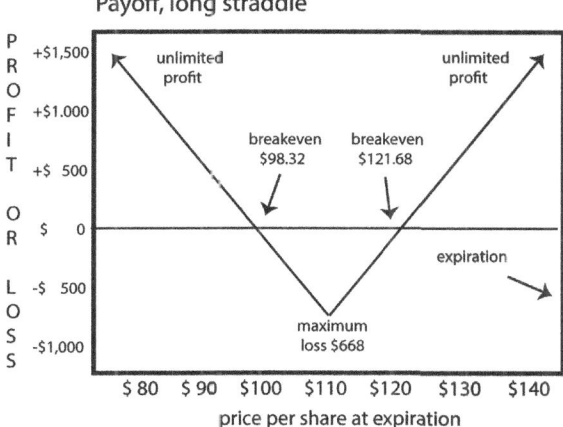

Fig. 8.1 Payoff, long straddle—prepared by the author

Short Straddles

A trader in long straddles must overcomes the cost of both call and put, and exceed that cost to generate profit. This is the disadvantage to a long straddle. In comparison, a short straddle (also called a sell straddle) provides significantly higher exercise risk but a greater likelihood of profits, assuming historical volatility remains relatively low. For long straddles used in a swing trading approach, the chances for profit are quite narrow:

> When trading options on a daily basis, it is difficult to gain a profit due to wide bid-ask spreads involved in option pricing; i.e., the change of the underlying asset is not large enough to cover the bid-ask spread of the option.[9]

However, the potential profits from short straddles are much more promising. One study compared long and short returns and concluded that:

> … for very short-term and short-term options, straddle returns decrease from deep out-of-the-money through deep in-the-money classes, thereby changing sign from plus to minus. Strategies with at-the-money calls and puts in these term classes yield significantly negative returns of approximately -1.0 and -0.5 per cent per day, respectively. Selling straddles created from short-term

---

[9]The example for lower loss is based on assumed decrease of 25 points in the underlying.

at-the-money options thus promise remarkable (excess) returns that could easily cover usual transaction costs encountered in the option market.[10]

The disparity in both risk and potential between long and short straddles represents extremes in the risk/return of each side. Construction of this strategy requires the opening of an equal number of short calls and short puts with the same strike and expiration. This strategy offers a limited profit, achieved when the underlying price is exactly at the strike of the two straddle sides:

$$P-F = M$$

- P premium received
- F trading fees
- M maximum profit

Breakeven occurs at two points, one above and the other below the strike:

$$S + (P-F) = B_u$$
$$S - (P-F) = B_l$$

- S strike
- P premium received
- F trading fees
- B_u upper breakeven
- B_l lower breakeven

The maximum loss is unlimited in either direction. (As previously noted, the lower loss is actually limited because the underlying price cannot decline below zero in the worst case.). If the underlying moves strongly beyond the breakeven range, losses are incurred:

Upper loss occurs when: $U > S - (P - F)$
Lower loss occurs when: $U < S - (P - F)$

- U underlying
- S strike
- P premium received
- F trading fees

---

[10] The example for upper profit is based on assumed increase of 15 points in the underlying.

8 Strategic Payoff: Straddles

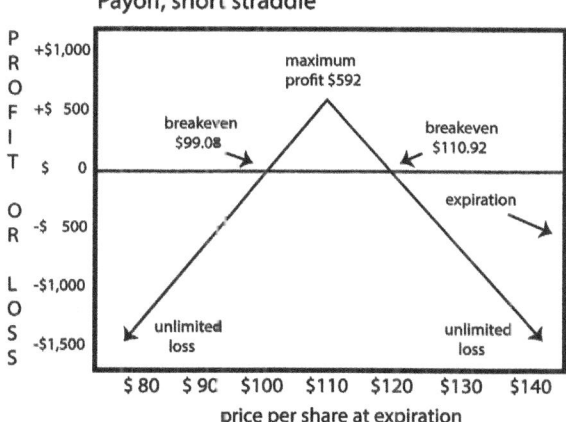

Fig. 8.2 Payoff, short straddle—prepared by the author

The maximum loss is equal to one of two outcomes:

$$U - S - (P - F) = M_u$$
$$S - U - (P - F) = M_l$$

- U underlying
- S strike
- P premium received
- F trading fees
- M_u maximum upper loss
- M_l maximum lower loss

Referring to Table 8.1, a short straddle can be constructed with the following ATM options:

Sell 105 call, bid 3.20, less trading fees = −$311
Sell 105 put, bid 2.90, less trading fees = −$281
Net credit = −$592

The outcomes for this trade are:

Maximum profit: $610−$18 = $592
Upper breakeven: 105+5.92 = 110.92 ($110.92)
Lower breakeven: 105−5.92−99.08 ($99.08)

Upper loss: 112–05–5.92 = 1.08 ($108)[3]
Lower loss: 105–97–5.92 = 2.08 ($208)[4]

These outcomes are also illustrated in Fig. 8.2.

Covered Straddles

The high exercise risk of the short straddle can be resolved by covering the call side. The difference in risk between covered and uncovered calls is the core issue, thus presenting the covered straddle as a solution. This is a short straddle with the singular distinction that the call is covered via ownership of 100 shares of the underlying for each short call opened.

The short put side of the trade is not a problem. The market risk of the uncovered put is identical to the market risk of a covered call (see Chap. 6). The conversion of the short call solves the exercise risk issue as well as the volatility problems associated with shorting options, especially in shorting two separate options with the same strike. As a hedge, the covered straddle is an attractive strategy for hedging equity positions:

> The ever-increasing volatility in the financial markets throughout the world points towards an increasing need for hedging investment portfolios. Today, a number of specialized instruments are available that allow market participants to protect themselves against adverse market movements. However, the costs, inflexibilities, or the uncertainties of cash outflows render futures contracts and short selling unattractive, especially at the retail level. The inherent flexibility and certainties involved in the cash outflows make options contracts the most favored instruments for hedging purposes. [11]

The well-known risks and opportunities of the covered call – representing one-half of the covered straddle—are modified by inclusion of the put, making the call itself optimal in two respects: First, the volatility risk premium presents an attractive level of revenue and second, through cover, the otherwise high-risk strategy becomes conservative:

> The short straddle component is short volatility, but it also includes additional risk owing to its options' dynamic equity exposure … The short straddle's positive performance is a result of options' tendency to be richly priced – the volatility risk premium. The covered call is simply a portfolio that combines

---

[11]The example for lower profit is based on assumed decrease of 20 points in the underlying.

these two risk premiums, and the strategy's expected returns and risk are best viewed through this lens. More importantly, portfolio construction should be guided by directly targeting exposure to these two sources of return.[12]

By definition, a covered straddle is a short straddle—selling an equal number of calls and puts with the same strike and expiration—with the additional ownership of 100 shares of the underlying. This modifies the calculation of profit, breakeven and loss. Because the uncovered put is identical in terms of market risk to the covered call, the construction of the covered straddle is similar to writing two covered calls (with the notable exception that only 100 shares per call are required in the covered straddle, versus 200 shares for the covered call).

Maximum profit for the covered straddle occurs by expiration if and when the underlying trades at or above the strike, modified by the price of the underlying and by the net premium received:

$$(P - F) + S - U_b = M$$

P premium received
F trading fees
S strike
U_b basis in underlying
M maximum profit

Breakeven for the covered straddle is located at the underlying price plus the strike, minus net premium received:

$$(U_b + S - (P - F)) \div 2 = B$$

U_b basis in underlying
S strike
P premium received
F trading fees
B breakeven

Loss is unlimited, but this has to be qualified in terms of actual risks. Since the short call is covered, loss exposure has to be considered as conservative as long as the basis in the underlying is lower than the strike of this position. The uncovered put has the same risk posture as a covered call. As a result, the maximum loss is equivalent to a loss on two covered calls, even though the position is offset by only 100 shares (assuming one

---

[12] The example for upper loss is based on assumed increase of 15 points in the underlying.

each of a short call and short put). In addition, the "unlimited" loss is in fact limited since an underlying price can only decline to zero in the worst case.

With these qualifiers in mind, maximum loss occurs when the underlying current price falls below the net of the underlying purchase price plus the net premium received. In this situation, the short call expires worthless and the short put is in the money. Thus, the put is exercised or has to be bought to close at a loss. The maximum loss is calculated as:

$$(U_b - U_c) + (P_c - P_b) = M$$

U_b underlying basis price
U_c underlying current price
P_c put, current premium
P_b put, net sale premium
M maximum loss

In this formula, the overall loss consists of three components: loss on the stock, loss on the put, and gain from original option premium received. However, this implies that the loss would be *realized* and that is not always the case. The loss on stock remains a paper loss unless shares are sold. The loss on the put is realized only if it is bought to close or exercised, but in practice it can be rolled forward to avoid exercise.

A covered straddle can be set up based on the positions in Table 8.1. The assumption has to include the purchase of shares. Assuming that occurred previously at the price of $102 per share (in other words, the trader already owns shares and expands the position by creating the straddle), the covered straddle with options at the money consists of:

100 shares @ $102 per share, $10,200 plus trading fees = $10,209
Sell 105 call, bid 3.20, less trading fees = $311
Sell 105 put, bid 2.90, less trading fees = $281
    Net equity long position = $10,209
    Net short options position = −$592

The outcomes for this trade are:
Maximum profit: 5.92 + 105−102 = 8.92 ($892)
Breakeven: (102 + 105−5.92) ÷ 2 = 100.54 ($100.54)
Maximum loss: (102−92) + (10−2.81) = 17.19 ($1,719)[5]

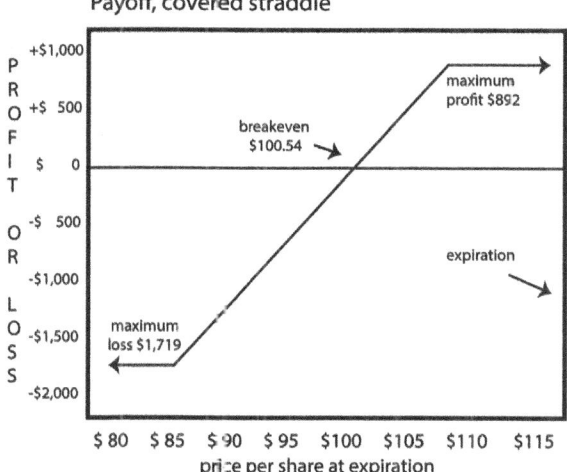

Fig. 8.3 Payoff, covered straddle—prepared by the author

This maximum loss (worst case) outcome is illustrated in Fig. 8.3.

This worst-case analysis cannot be relied upon by the trader who is aware of alternatives. The stock does not have to be sold to realize the maximum loss, and the short put's exercise can be avoided by rolling forward. That put can also be covered with the purchase of a second put, which either offsets the loss or mitigates its impact. However, as a comparative analysis, the outcomes for a covered straddle should be compared (in terms of opportunity as well as risk) on the same basis as other strategies.

Risk and Payoff Calculations: Strangles

The *strangle* modifies the straddle by setting up OTM positions in place at ATM. This completely changes the potential outcomes for both long and short set-ups, and may present a more realistic form of strategies hedging. The straddle presented in the previous section appears to present a relative long shot for profitable outcomes. The strangle solves this problem, at least to a degree.

Even so, the well-known risks of both long and short strategies cannot be overcome completely. A long strangle requires considerable price movement on one side or the other; and although the strangle is less expensive than the long straddle, it requires more underlying movement to overcome the OTM status of each option. The short strangle provides lower exercise risk but also provides less premium income, so that the profitable zone is also reduced due to the OTM status of options. The short strangle may provide an appropriate level of strategy when the underlying is trading in a well-defined consolidation range. During this period, a breakout and beginning of a dynamic trading range, once identified, is the trigger for closing one side of the short strangle (the side moving toward the money), whereas the remaining side can be left to expire or closed once time value has evaporated.

Long Strangles

The long strangle (also called a "buy strangle") combines a long call and a long put at the same expiration date, both out of the money. It sets up the potential for unlimited profit, with risk limited to the cost of the premium for both options. It is most advantageous when volatility is high or expected to trend to higher levels.

A profit is achieved when one side or the other moves far enough to exceed the cost for opening the position, and may occur either above the strike range or below. The formula:

$$U - S_c - (P + F) = P_u$$
$$S_p - U - (P + F) = P_l$$

U underlying price
S_c call strike
S_p put strike
P premium paid
F trading fees
P_u upper profit
P_l lower profit

Breakeven also occurs on either side of the strangle position, upper and lower.

The formula:

$$S_c + (P + F) = B_u$$
$$S_p - (P + F) = B_u$$

S_c call strike
S_p put strike
P premium paid

Table 8.2 Calls and puts, Alphabet (GOOG)

Strike [a]	Calls bid ask		Puts bid ask	
12/23/16	19.00	20.90	3.60	4.20
780	15.50	17.00	4.90	5.50
785	12.30	13.50	6.40	7.20
790	9.50	19.50	8.40	9.40
795	7.20	8.00	10.90	12.00
800	5.10	5.90	13.80	15.20
805	3.40	4.30	17.30	18.80
810				

[a] GOOG's price at this time was $796.10, at the close on December 13, 2016

F trading fees
B_u upper breakeven
B_l lower breakeven

The risk of loss in the long strangle is limited to the initial debit paid for the long strangle. This occurs when price remains in between the two strikes by expiration, when neither option is sold to mitigate the overall loss. The formula:

$$P + F = M$$

P premium paid
F trading fees
M maximum loss

A long strangle can be set up with options as shown on Table 8.2. Following are the entries to create a long strangle, using OTM options:

Buy 800 call, ask 8.00, plus trading fees = 8.09
Buy 790 put, ask 7.20, plus trading fees = 7.29
Total debit = 15.38 ($1538)

The challenge to any trader entering this position is apparent. In order to exceed breakeven, the underlying must exceed 15.38 points before expiration. Applying the outcome formulas to this strangle:

Upper profit: 820−800−15.38 = 4.62[6]
Lower profit: 790−765−15.38 = 9.62[7]
Upper breakeven: 820+15.38 = 835.38 ($835.38)

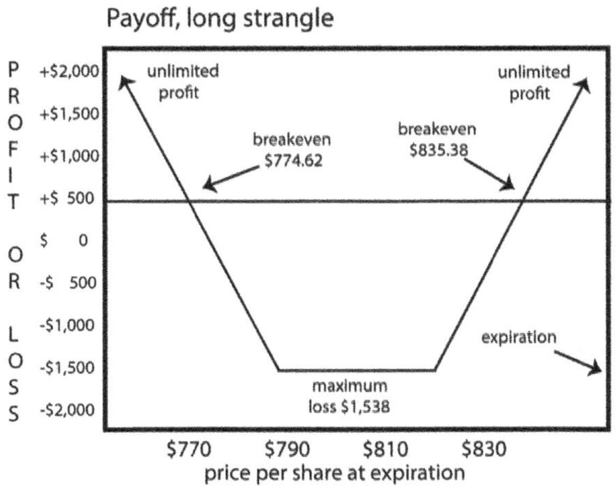

Payoff, long strangle

```
PROFIT OR LOSS
+$2,000  unlimited profit                    unlimited profit
+$1,500
+$1,000       breakeven        breakeven
              $774.62          $835.38
+$ 500
$ 0
-$ 500
-$1,000
-$1,500                                      expiration
         maximum
-$2,000  loss $1,538
         $770   $790   $810   $830
         price per share at expiration
```

Fig. 8.4 Payoff, long strangle—prepared by the author

Lower breakeven: 790−15.38 = 774.62 ($774.62)
Maximum loss: 15.20+0.18 = 15.38 ($1538)

This range of outcomes is illustrated in the diagram on Fig. 8.4.

Short Strangles

The long strangle requires high volatility in order to accomplish the strong price movement in the underlying. In comparison, a short strangle (also called a "sell strangle") consists of OTM short calls and puts, and will perform best in times of low volatility. This strategy offers a limited profit with unlimited risk, and is set up as a net credit.

Maximum profit is limited to the credit received:

$$P - F = M$$

P premium received
F trading fees
M maximum profit

Breakeven occurs in two positions, one above the call strike and one below the put strike:

$$S_c + (P - F) = B_u$$
$$S_p - (P - F) = B_l$$

S_c call strike
S_p put strike
P premium received
F trading fees
B_u upper breakeven
B_l lower breakeven

Loss is unlimited once the breakeven levels are passed. A maximum loss occurs either above or below the strike range:

$$U - S_c - (P - F) = M_u$$
$$S_p - U - (P - F) = M_l$$

U underlying price
S_c call strike
S_p put strike
P premium received
F trading fees
M_u upper maximum loss
M_l lower maximum loss

Employing the same strikes as those for the long strangle, a short strangle can be established based on the options summarized in Table 8.2:

Sell 800 call, bid 7.20, less trading fees = −7.11
Sell 790 put, bid 6.40, less trading fees = −6.31
Net credit = −13.42 (−$1342)

Outcomes are realized based on these positions as:

Maximum profit: 13.60–0.18 = 13.42 ($1342)
Upper breakeven: 800+13.42 = 813.42 ($813.42)
Lower breakeven: 790–13.42 = 776.58 ($776.58)
Upper loss: 820–800–13.42 = 6.58 ($658)[8]
Lower loss: 790–765 − 13.42 = 11.58 ($1158)[9]

The range of outcomes is summarized in Fig. 8.5.

As with all instances involving short puts as segments of a strategy, the maximum loss is not truly "unlimited" even though that description

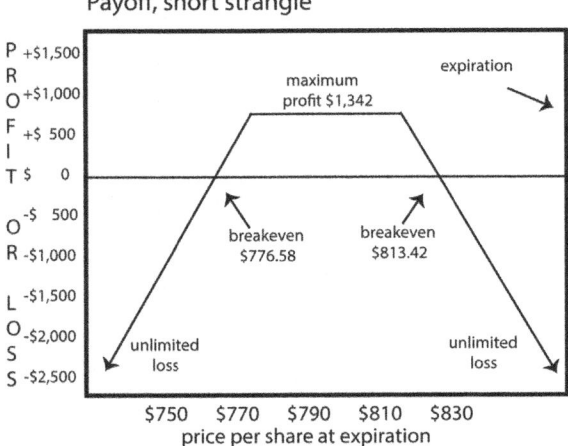

Fig. 8.5 Payoff, short strangle—prepared by the author

defines the risk potential existing on both sides. In practice, the downside loss is limited because the underlying cannot decline below a value of zero.

Most descriptions of the short strangle suggest the most likely strikes should be OTM but as close as possible to the current underlying price. This is a sensible strategic limitation in some circumstances; however, given the unlimited loss potential, there are instances in which a strike buffer zone also makes sense. In the case of GOOG used in the preceding examples, the stock chart as of December 13 reveals a consolidation trend in effect for over four months (Fig. 8.6).

When consolidation is the current pattern, timing for a short strangle is ideal. Based on the Bollinger bandwidth of 48.67 points, a 30-point distance between call and put strikes (each 15 points away from the price of the underlying) sets up a comfortable buffer zone for a short straddle. Using the listings in Table 8.2, the following positions create a very low-risk short straddle:

Sell 810 call, bid 3.40, less trading fees = −$331
Sell 780 put, bid 3.60, less trading fees = −$351
Net credit = −6.82 (−$682)

In this example, relying on the historical volatility displayed by the Bollinger upper and lower bands provides guidance for establishing a buffer zone. As long as the price levels are range-bound in a current consolidation trend,

Fig. 8.6 GOOG price chart—chart courtesy of StockCharts.com

this method reduces market risk considerably, especially for very short-term options. The example involves options expiring in only 10 days, the idela timing to exploit time decay.

Long Gut Strangle

A variation on the strangle is called the *gut strangle*. In this version, the OTM options are replaced with ITM contracts. This sets up the potential for unlimited profits and limited losses, and is designed to perform best during times of high. In this regard, the ITM gut positions perform in the same manner as the OTM long strangles; however, movement in the underlying must be strong enough to overcome the cost of the position, which will be significantly higher than for the OTM strangle strategy. Rationale for the higher cost is that with ITM positions, a strong directional move becomes profitable rapidly in comparison to the OTM strangle options.

Profit is unlimited in the gut strangle, and occurs when either the long call or long put moves farther than the overall net debit of the position. The profit formula:

$$U - S_c - (P + F) = P_u$$
$$S_p - U - (P + FP) = P_l$$

U underlying price
S_c call strike
S_p put strike

P   premium paid
F   trading fees
P_u upper profit
P_l lower profit

Breakeven also occurs at two points, one above the call strike and one below the put:

$$S_c + (P - F) = B_u$$
$$S_p - (P - F) = B_l$$

S_c call strike
S_p put strike
P   premium received
F   trading fees
B_u upper breakeven
B_l lower breakeven

Maximum loss occurs when the underlying finishes up in between the two strikes, and is the net difference between the two:

$$(P + F) + (S_p - S_c) = M$$

P   premium paid
F   trading fees
S_p put strike
S_c call strike
M   maximum loss

A long gut strangle can be set up based on the options shown in Table 8.3. For example:

Buy 60 call, ask 4.30, plus trading fees = 4.39
Buy 70 put, ask 7.00, plus trading fees = 7.09
Total debit = 11.48 ($1148)

The outcomes:

Upper profit: 85–70–11.48 = 3.52 ($352)[10]
Lower profit: 60–40–11.48 = 8.52 ($852)[11]

Table 8.3 Calls and puts, General Mills (GIS)

Strike [a]	Calls bid ask		Puts bid ask	
1/20/17				
55	8.55	9.10	0.15	0.24
60	4.10	4.30	0.59	0.71
65	1.02	1.11	2.65	2.77
70	0.17	0.27	6.75	7.00

[a] GIS's price at this time was $63.74, at the close on December 13, 2016

Upper breakeven: 70+11.48 = 81.48 ($81.48)
Lower breakeven: 60−11.48 = 43.52 ($48.52)
Maximum loss: 11.48 + (60−70) = 1.48 ($148)

$$(P + F) + (S_p - S_c) = M$$

P premium paid
F trading fees
S_p put strike
S_c call strike
M maximum loss

This outcome is illustrated in Fig. 8.7.

Short Gut Strangle

The short gut trade is the opposite of the long; it is composed of an equal number of short calls and short puts with the same expiration, both in the money. Profit is limited by the level of net credit received, and potential loss is unlimited. The strategy is maximized when volatility in the underlying is relatively low.

Maximum profit occurs when the underlying is trading in between the two short strikes. Because both are in the money, the advantage is the loss of time value. Accordingly, the analysis at expiration is not realistic because it does not include computation of the inevitable exercise of both sides. In practice, both options would be closed at a profit to exploit time decay while avoiding exercise. The profit calculation:

$$(P - F) + S_p - S_c = M$$

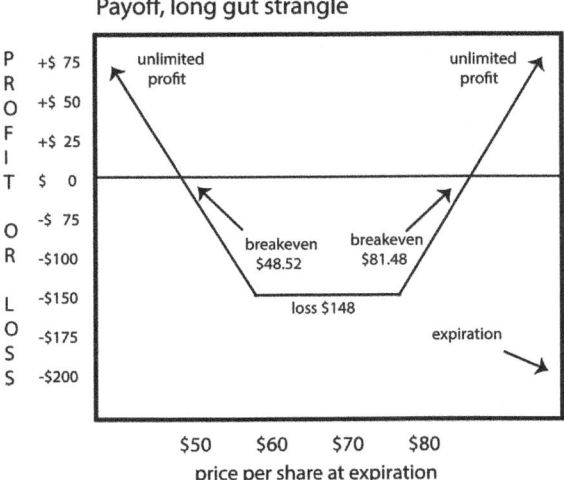

Fig. 8.7 Payoff, long gut strangle—prepared by the author

P   premium received
F   trading fees paid
S_p put strike
S_c call strike
M   maximum profit

A breakeven occurs in two positions, one above the call's strike and the other below the put's strike:

$$(P - F) + S_c = B_u$$
$$S_p - (P - F) = B_l$$

P   premium received
F   trading fees
S_c call strike
S_p put strike
B_u upper breakeven
B_l lower breakeven

Net loss is unlimited, and occurs whenever the underlying moves strongly in one direction or the other beyond the net of strikes adjusted by net premium received. The "unlimited" downside maximum loss is limited because price may not decline below zero; however, the use of the term "unlimited" is used to demonstrate contrast with a limited loss. The formula:

$$U - S_c - (P - F) = M_u$$
$$S_p - U - (P - F) = M_l$$

U underlying
S_c call strike
S_p put strike
P premium received
F trading fees
M_u maximum upper loss
M_l maximum lower loss

A short gut strangle can be established based on Table 8.3, for General Mills (GIS) options, with the following:

Sell 60 call, bid 4.10, less trading fees = −$401
Sell 70 put, bid 6.75, less trading fees = −$666
Total credit −$1067

The range of outcomes:

Maximum profit: 10.67+(70–60) = 20.67 ($2067)
Upper breakeven: 10.67+60 = 70.67 ($70.67)
Lower breakeven: 70–10.67 = 59.33 ($59.33)
Upper loss: 75–60–10.67 = 4.33 ($433)[12]
Lower loss: 70–50–10.67 = 9.33 ($933)[13]

These outcomes are summarized in Fig. 8.8.

Risk and Payoff Calculations: Strips and Straps

A variation on the straddle is adding a ratio to one side or the other. Depending on how the strip or strap is set up, it may be clearly bearish or bullish.

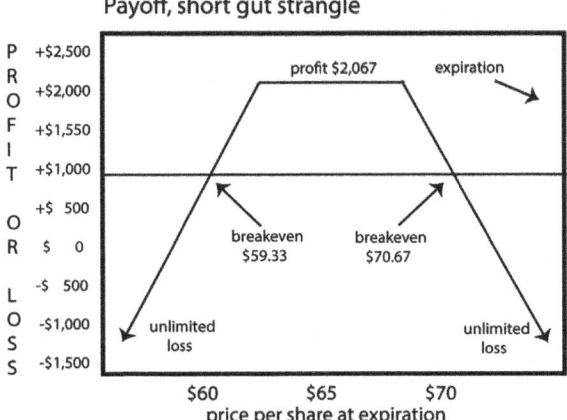

Fig. 8.8 Payoff, short gut strangle—prepared by the author

Strip

The *long strip* (also called "strip straddle") is a bearish variation of the better-known straddle, in which more puts are bought than calls. It remains a long position with a ratio of puts over calls, but sets up what many traders consider a desirable adjustment. The use of long options only reduces the risks associated with short options.

If the long strip price declines, the position becomes profitable, and more so than the one-to-one straddle, due to the higher number of long puts. However, because several long options are involved, the required underlying price movement must be considerable. This means that the long strip works best when volatility is high. As a swing trade, for example, a long strip often is entered when the trader believes a downside breakout is likely to occur, or when earnings are about to be announced, and a negative earnings surprise is expected. The direction of an earnings surprise is by no means an easy forecast to make, but in those cases when it does occur, the long strip would exploit a considerable move in the underlying.

Using specific options strategies to time trades based on expected earnings surprises is an unreliable timing mechanism, for several reasons. The reaction to earnings among small traders (individuals) is not likely to mirror behavior among large traders (institutions). In fact, the retail trader tends to overreact when a series of earnings surprises in the same direction has occurred. The assumption of a pending repeat of the same pattern may lead traders to proceed with exaggerated expectations. One study concluded:

... small traders' reaction relative to large traders becomes more negative over a series of negative surprises ... Similarly, results show that small traders' relative trade reaction becomes generally more positive for a series of positive surprises ...[13]

Yet another factor distorting the reliability of the earnings surprise itself is a tendency among managers to attempt to influence the analysts' estimates of reported revenues and earnings. Although the Sarbanes-Oxley Act of 2002 (SOX) was supposed to mitigate management's manipulation and distortion of reported earnings (notably what occurs between management and auditors), the same level of influence over analysts is not specifically addressed in the same manner. In 2002, a paper on this topic noted that:

... managers of some firms place great importance on meeting or exceeding analysts' expectations, which they achieve either by using their discretion over reported earnings (earnings management) or by guiding analysts' earnings forecasts downward to improve their firm's chances of meeting or beating the forecast when earnings are announced (forecast guidance).[14]

This observation was published in 2002, the same year the Sarbanes-Oxley Act (SOX) was passed. It might be assumed that the problem of manipulation over expectations no longer exists. However, SOX focused on the relationship between management and auditors, but did not affect what analysts report or predict. The problem persists to this day:

A federal rule bars companies from selectively disclosing material nonpublic information but doesn't prohibit private conversations in which companies can gently push analysts in helpful directions ... Some analysts, investor-relations officials, securities lawyers and executives say the signals have become so commonplace that the all-important question of whether a company beat estimates is more about theatrics than reality.[15]

With a cautionary observation for those relying on the consistency of repetitive earnings surprises, strategies such as the long strip may perform well in some circumstances, while in others the underlying might not perform as hoped, even in light of a significant earnings surprise.

---

[13]The example for lower loss is based on assumed decrease of 20 points in the underlying.
[14]The example for upper profit is based on assumed increase of 30 points in the underlying.
[15]The example for lower loss is based on assumed decrease of 10 points in the underlying.

The potential profit from the strip will be greater than that of the one-to-one option characteristics of a long straddle. A downward move in the underlying will create a multiplying affect when the number of puts exceeds the number of calls. In its most basic form, a strip consists of opening one long call and two long puts, at the same strike and expiration. This sets up the potential for unlimited downside profits. Profits are realized when the underlying price moves substantially away from the single strike; this may occur on either side of the strike. The formula for profit in a 2-put, 1-call strip is:

$$U - S - (P + F) = M_u$$
$$(S * 2) - U - (P + F) = M_l$$

U underlying price
S strike
P premium paid
F trading fees
M_u maximum upper profit
M_l maximum lower profit

Breakeven also occurs at two price points:

$$S + (P + F) = B_u$$
$$S - (P + F) = B_l$$

S strike
P premium paid
F trading fees
B_u upper breakeven
B_l lower breakeven

Maximum net loss occurs when the underlying price is exactly at the strike of the position. It is limited to the cost of the trade:

$$P + F = M$$

P premium paid
F trading fees
M maximum loss

Table 8.4 Calls and puts, Intuitive Surgical (ISRG)

Strike [a]	Calls bid ask		Puts bid ask	
12/23/16	18.50	21.50	2.70	3.40
620	14.80	16.00	2.00	5.20
625	11.30	12.50	5.20	6.20
630	8.00	11.20	7.00	8.20
635	6.20	7.00	8.00	11.50
640	4.30	5.00	12.70	14.00
645	3.00	3.50	16.10	18.00
650				

[a] ISRG's price at this time was $635.97, at the close on December 13, 2016

For example, a long strip can be set up with options shown on Table 8.4 A 2-to-1 strip may consist of:

Buy two 635 puts, ask 8.20, plus trading fees = 16.50
Buy one 635 call, ask 11.20, plus trading fees = 11.29
Total debit = 27.79 ($2779)

This position creates outcomes of:

Maximum upper profit: 665–635 – 27.79 = 2.21 ($221)[14]
Maximum lower profit: (635 * 2)–625–27.79 = 617.21 ($617.21[15])
Upper breakeven: 635+27.79 = 662.79 ($662.79)
Lower breakeven: 635–27.79 = 607.21 ($607.21)
Maximum loss: 2760+19 = 2779 ($2779)

These outcomes are summarized in the graph in Fig. 8.9.

A short strip can also be opened, with all of the positions reversed to reflect short rather than long. More puts are sold than calls, setting up a net debit. The potential profit is limited to the overall net credit, while potential losses are unlimited. Losses on the downside accumulate at an accelerated rate. For example, with two short puts, losses will equal two points for each point of decline in the underlying. The payoff summary is also flipped between potential profits and losses.

An analysis of the limited profit (the net credit received) should be compared to the risk of having short options exposed. This strategy relies on time decay on both sides, or requires early close or rolling to avoid exercise.

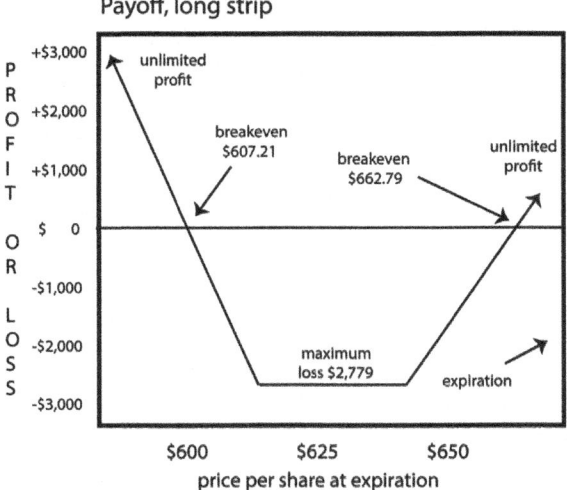

Fig. 8.9 Payoff, long strip—prepared by the author

Strap

The *long strap* (also called a "strap straddle") is a bullish variation on the long straddle with equal numbers of options. A long strap consists of a greater number of long calls. In a 2-to-1 position, two long calls are opened along with one long put, at the same strike and expiration.

The long strap is the opposite of the long strip, in the sense that it is bullish rather than bearish, and also because there are a greater number of calls than puts.

Profits are greater on the upside, but both directions hold the potential for unlimited profits. The formula:

$$2 * (U - S) - (P + F) = M_u$$
$$S - U - (P + F) = M_l$$

- U underlying
- S strike
- P premium paid
- F trading fees
- M_u maximum upper profit
- M_l maximum lower profit

Breakeven occurs at two points:

$$S + (P + F) = B_u$$
$$S - (P + F) = B_l$$

S   strike
P   premium paid
F   trading fees
B_u upper breakeven
B_l lower breakeven

Maximum loss is equal to the overall cost to open the long strap:

$$P + F = M$$

P   premium paid
F   trading fees
M   maximum loss

A long strap can be created using options shown on previously introduced Table 8.4. For example:

Buy two 635 calls, ask 11.20, plus trading fees = 22.50
Buy one 635 put, ask 8.20, plus trading fees = 8.29
Total debit = 30.79 ($3079)

The outcomes for this trade are:

Upper profit: 2 * (655−635) − 30.79 = 9.21 ($921)[16]
Lower profit: 635−600−30.79 = 4.21 ($421)[17]
Upper breakeven: 635+30.79 = 665/79 ($665.79)
Lower breakeven: 635−30.79 = 604.21 ($604.21)
Maximum loss: 30.60+0.19 = 30.79 ($3079)

Figure 8.10 summarizes these outcomes.

A short strap is the opposite of the long; it combines a greater number of short calls than short puts. As a combined short strategy, exercise risks

---

[16] The example for upper profit is based on assumed increase of 20 points in the underlying.
[17] The example for lower profit is based on assumed decrease of 30 points in the underlying.

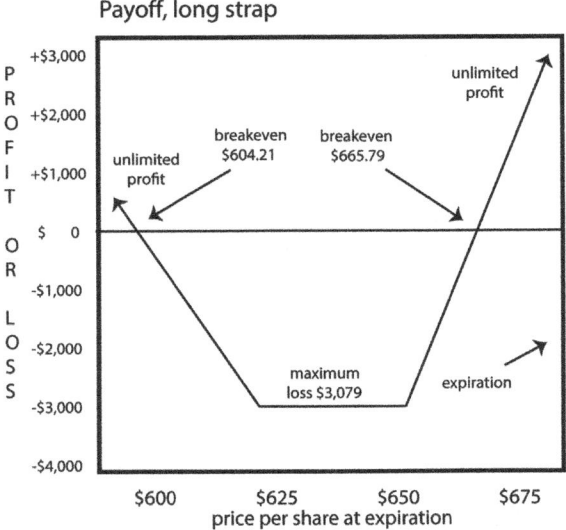

Fig. 8.10 Payoff, long strap—prepared by the author

have to be considered at all times except when the underlying price is identical to the strike. The collateral requirements for uncovered short options is another factor to keep in mind when assessing the profit potential versus the risks of a complex short position. Unlike the long strap, a short strategy is not a hedge but a speculative device intended to exploit rapid time decay. The ideal outcome is to experience low volatility so that both sides can be closed profitably. However, because the short strap usually is described as set up with ATM options, no buffer is included to absorb unexpected underlying price movement. As a result, this strategy should be avoided during periods in which earnings are reported, as an earnings surprise could create a severe ITM result. Months in which ex-dividend date occurs should also be avoided with the short strap, as this is the most likely timing for early exercise of ITM calls.

The behavior profile for a short strap is similar to that of a long strip; however, with open short positions rather than long, the offsetting net credit advantage buffers risk, whereas the corresponding long strip limits risk. The same argument applies to the short strip with a greater number of short puts, in comparison to the long strap. Here as well, the use of short positions (strip) buffers the loss range but the use of long positions (strap) limits maximum loss.

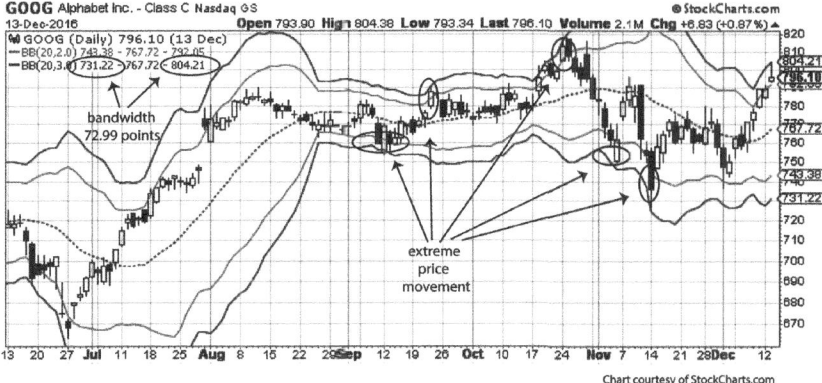

Fig. 8.11 Alphabet (GOOG) with BB expansion—chart courtesy of StockCharts.com

Strategic Selection of Strikes

For all straddles and variations on straddles, selection of strikes is one element of risk management. This is especially important for positions such as the short strangle, when exercise risk is best handled by setting up a buffer zone out of the money on both sides. The problem with this, however, is that for lower-priced underlying securities, a buffer zone often is accompanied by such a low premium level that the position is not worth the exposure.

For this reason, a short-term (one to two week) term to expiration is advantageous for short strangle and similar trades. At the same time, higher-priced underlying securities are more likely to provide the acceptable range of premium to justify the trade. To determine whether a trade is viable with a buffer zone between OTM strikes and the current price, add a new layer of Bollinger Band analysis. Figure 8.11 repeats the previously introduced chart for Alphabet (GOOG) with the addition of a revised Bollinger Band summary.

This BB expansion combines the default of two standard deviations (with band width of 48.67 points, and the new three standard deviations (outer rings) with bandwidth of 72.99 points. The ideal placement of strikes based on the "probability matrix" of Bollinger Bands is any strike between the two- and three-standard deviation ranges.

Price does move outside of the default two-standard deviation BB range. However, on this chart, price never closed outside of the wider three-standard deviation range. The extreme price movements (marked on the chart) pointed out consistent reversal points on the chart. When price touched or moved close to the outer bands, it reversed without fail. Using this addition to a chart points to a reliable and consistent signal for timing of short trades. At the bottom band, short puts are ideal because price is expected to reverse and move back into the range of the two-standard deviation area. At the top, the same argument can be made for timing of short calls.

Using the combined two-and-three BB system, the opening of short options is not limited to entering both at the same time. It can be offset with an alternative strategy that is similar to a short strangle, but with superior timing. Thus, a short put is opened at the lower extreme price movement touching the lower band; and the short call is opened at the top. The staggered timing of each side also points to the timing for a buy to close order in both directions.

The selection of a strike may also be timed for ATM positions at the time of extreme price movement. This solves the problem for the short strangle writer. When opening both sides when the underlying is at mid-range, the extreme OTM positions will be very cheap, and justifying the position limits activity to higher-price underlying securities. However, timing ATM trades for the extreme price movement reduces risks while facilitating a strangle-like trade based on likely reversal points.

Trying a similar trading strategy with the use of the default two-standard deviation BB is flawed because price often exceeds the outer bands, at times for several sessions. Price does eventually retreat back into BB range, but most traders will acknowledge that opening uncovered ATM short positions is high-risk even when timed for the violation of two-standard deviation Bollinger movement.

Combining two and three standard deviations on a chart is an effective swing trading approach to solve the short option position. The strangle is the most problematic as long as a trader limits the strategy to opening both sides at the same time. When doing so, one side or the other is always at risk. This risk is aggravated by the tendency to select strikes close enough to the money to justify the risk with attractive premium. The alternative of staggering the timing of entry for each side exploits the cyclical movement of the underlying while maximizing premium income.

This staggered entry and exit is closely associated with swing trading, but a problem with this system is that it normally relies on identification of strong reversal and confirmation signals. These are known to fail on occasion, so the use of the Bollinger Band system is superior. However, it is not limited to swing trading or similar speculative methods. The combined BB method is also effective as part of a recovery strategy. Every options trader has experienced times of loss, notably when combining options with equities. For example, owning 100 shares and writing a covered call often is described as a low-risk trade. However, if the underlying price declines far enough, it creates a paper loss. Traders may wait out the price range, hoping for reversal. But that could take weeks or even months, and might never occur at all. At such times, the short strangle is one type of trade designed to recover from a loss on another trade. However, the tendency to take risks beyond a normal risk tolerance level may easily lead to even larger losses. At these times, the staggered approach to short options (similar to the strangle) is a sensible recovery strategy. No trade is foolproof, but this provides one of the best timing mechanisms, appropriate for traders willing to set up and monitor uncovered short positions.

In addition to purely speculative trades or recovery strategies, the staggered short works quite well in times of consolidation. Referring again to Fig. 8.11, Alphabet (GOOG) Alphabet (GOOG) traded in a narrow consolidation range between $760 and $790 for two months between early August and mid-October. In this period, two extreme price levels appeared, one in early September and the other immediately after in late September. These patterns provided exceptional timing for a short put (first case), following by a short call (second case). By opening these options as indicated within the consolidation period, risks were minimal. These short trades (ideally with expiration of two weeks or less) could be closed once time value declined substantially, or upon seeing the underlying price move to the opposite extreme. They could also be left to expire worthless; however, closing at a point when time value has mostly evaporated often makes more sense, not only to avoid exercise but also to free up collateral for the opposite end of the swing.

Even though straddles present risk elements whether long or short, they serve a purpose in many situations. As volatility trades, straddles, strangles, strips and stripes each provide specific opportunities as well as risks. By understanding how to calculate profit, breakeven and loss, traders are better equipped to critical evaluate a trade and to determine whether or not the balance between potential profit and loss justifies the position.

Chapter Summary.

Straddles are best timed for entry at high volatility (long) or low volatility (short)
The straddle's appeal is in its combination of bullish and bearish sentiment
Covered straddles vastly reduce short-side risk by covering the call side
Strangles reduce exercise risk through selection of strikes away from the money

Strips and straps are opposites in terms call/put selection as well as long/short

Strategic selection and timing of strikes is an effective risk-reduction concept.

Notes

1. Enke, David & Sunisa Amornwattana (2008). A hybrid derivative trading system based on volatility and return forecasting. *The Engineering Economist, 53*(3), 259–292.
2. *Farlex Financial Dictionary.* (2009). Retrieved December 13 2016 from http://financial-dictionary.thefreedictionary.com/double+option.
3. Chaput, J. & Louis H. Ederington (Summer, 2003). Option spread and combination trading. *Journal of Derivatives.* 10(4):70–88.
4. Mello, A. S. & Henrik J. Neuhaus (1998). A portfolio approach to risk reduction in discretely rebalanced option hedges. *Management Science, 44*(7), 921–934.
5. Jorion, Philippe (June 1995). Predicting volatility in the foreign exchange market. *The Journal of Finance, 50*(2), 50.
6. Natenberg, Sheldon (1994). *Option Volatility and Pricing.* New York: McGraw-Hill. p. 292.
7. Goltz, Felix & Wan Ni Lai (2009). Empirical properties of straddle returns. *Journal of Derivatives, 17*(1), 38–48, 4–5.
8. Chaput & Ederington, *Op;. Cit.* 243–279.
9. Enke & Amornwattana, *Op. Cit.*
10. Wilkens, Sascha (2007). Option returns versus asset-pricing theory: Evidence from the European option market. *Journal of Derivatives & Hedge Funds, 13*(2), 170–176.
11. Aggarwal, Navdeep & Mohit Gupta (2013). Portfolio hedging through options: Covered call versus protective put. *Journal of Management Research, 13*(2), 118–126.
12. Israelov, Roni & Lars N. Nielsen (2014). Covered call strategies: One fact and eight myths. *Financial Analysts Journal, 70*(6), 23–31.
13. Shanthikumar, Devin M. (2012). Consecutive earnings surprises: Small and large trader reactions. *The Accounting Review, 87*(5), 1709–1736.
14. Matsumoto, Dawn A. (2002). Management's incentives to avoid negative earnings surprises. *The Accounting Review, 77*(3), 483–514.
15. Grytam Thomas, Serena Ng & Theo Francis (August 4, 2016). Companies Routinely Steer Analysts to Deliver Earnings Surprises. *The Wall Street Journal.*

9

Probability and Risk

Chapter Objectives

- study the nature of skew and abnormal distribution
- compare the nature of fat tails to normal distribution
- articulate how the human element affects the risk universe
- equate risk theories with individual risk profile and tolerance
- analyze variance and its effect on risk perception
- fold the theory of fluency into analysis of random variables.

Options traders tend to focus on probability as part of the decision to enter (or exit) a particular strategy. They also analyze risk in the context of which positions are good fits for their risk profile. However, these two aspects of trading—probability and risk—are often considered as separate considerations, and may not be evaluated as two aspects of the same concern.

In studying the concepts of probability and risk, it is apparent that they are related and inseparable. Even so, theory and practice are not always placed together. Emphasis on the probability of an option finishing in or out of the money is one form of analysis. However, the set of questions not always asked at the same time is, "What is the risk of that probability not being realized and, consequently, what is the risk of a loss? How much would the loss be? Is that a reasonable risk to take?

For example, consider a situation where a short option is open on Wednesday, set to expire on Friday. However, the company is reporting earnings on Thursday. In the past, large earnings surprises have moved the underlying price in an exaggerated manner, only to self-correct within a few sessions. Given the fact that the current short position expires in 2 days, a *prudent* course would be to close on Wednesday and make a second decision on Friday (for example, opening a new short position to expire the following week). In this way, the risk of an earnings surprise and potential large loss is avoided. However, the probability analysis reveals only a very small chance of an ITM finish. Based on probability, the trader takes no action.

The possibility of an earnings surprise is high, and this is not always taken into account in determining the probability of how the option ends up in 2 days. This is especially the case when the short position is deep out of the money. To the casual observer, the chances of a big move and resulting loss are remote. Even so, it makes sense to combine the probability with a complete appreciation of the risks involved. Avoiding the unexpected is just as important as maximizing profit; a decision to reduce the possibility of the unexpected may be just as important, especially when that outcome would add unacceptable risks to the equation.

Options traders tend to rely on statistical possibilities, but calculation of probability relies on assumptions, and these are not always reliable. In most statistical analyses, normal distribution is assumed to apply to the analysis. However, with options trading, the likelihood of outcomes based on ending prices of the underlying and the option are not limited to a finite range, and as a consequence, probability and outcome are not subject to normal distribution.

Abnormal Distribution of Options Trading

The *skew* of probability distribution is a key factor in how probability is calculated. With a fixed population, distribution is straightforward. For example, a consumer is asked in a survey to choose preference among 10 products. There are only four different outcomes and expectation is that a strongly favored product will be revealed on a bell curve. The expectation would further assume the standard deviation of outcomes, so that 68% of outcomes would reside at the top portion equal to one standard deviation; and 95% would fall within two standard deviations.

In options probability, the skew represents the distortion of the bell curve, which is assumed to be symmetrical with fixed numbers of outcomes and samples. The skew is a visual representation of the random variable's distribution, and may be either positive or negative. In a positive skew, the tail on the right side is longer; and in a negative skew, the tail on the left side is longer. This is summarized in Fig. 9.1.

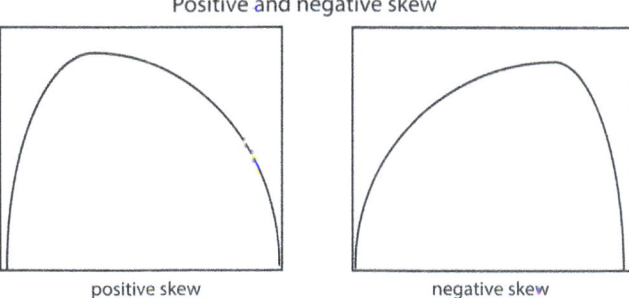

Fig. 9.1 Positive and negative skew—prepared by the author

This focus on where distribution lies is typical of probability distribution for options trading. It is not realistic to expect to find normal distribution because the outcomes are not fixed in number, and the sample of the outcome population cannot be defined in finite terms. A simplified example of skew can be observed by perceiving the likely normal distributed among five values, 10, 20, 30, 40 and 50. If these are normally distributed around the central value of 30, you expect to observe a perfect bell curve with no skew. However, if an additional value of 60 is added but distribution remains centered on 30, the distribution will be skewed as a consequence.

In calculation of skew, the mean, mode and median come into play. A calculation called Pearson's First Skewness Coefficient[1] (or mode skewness) is:

Pearson's First Skewness Coefficient

$$(A - MO) \div SD = S$$

A mean (average)
MO mode
SD standard deviation
S skewness coefficient

A second version is based not on mode, but on median. Pearson's Second Skewness Coefficient is calculated as:

Pearson's Second Skewness Coefficient

$$(A - ME) \div SD = S$$

A   mean (average)
ME median
SD standard deviation
S   skewness coefficient

The purpose of calculating skew—whether based on mode or median—is to identify the degree of distortion in probability distribution. Rather than the orderly bell curve expected with a finite number of likely outcomes, options probability is far more chaotic, and the greater the skew, the greater the risk. This is one method for equating probability and its abnormal appearance (skew) with actual degrees of risk.

The use of historical volatility as a means for quantifying probability is related directly to skew. Indicators like Bollinger Bands set up the probability matrix that may visually direct traders to articulate the risk, not only in a particular trade, but more specifically in the selection of strikes. The proximity between current underlying price and the selected strike defines the probability of high potential for profits (or losses).

Long positions with deep OTM strikes are cheaper, but contain lower probability of profitable outcome. Longer time periods are more expensive but set up greater potentials for the underlying price to move in a desired direction. Short positions contain the same considerations but with opposite significance. Employing Bollinger Bands to set up buffer zones between current price and option strikes (below the lower band or above the upper band) reduce the probability of ITM movement and exercise. However, the greater the distance between current price and option strike, the less premium a short seller receives. The longer the time to expiration, the higher the dollar value of premium, but the greater the duration of exposure. The shorter the expiration, the more rapid the time decay (which in turn increases the probability of being able to close profitably or allow to expire).

These risk and probability considerations are expressed in skew, just as the probability itself is defined within the historical volatility zone established by Bollinger Bands. Skewness is yet another visualization of the same issue, with a statistical focus. It can be calculated in many additional ways; the point to be remembered here is that probability distribution for options trading cannot be expected to be distributed normally. The outcome invariably reflects a fat tail on one side or the other in the distribution map.

historical volatility, consolidating range

Fig. 9.2 Historical volatility, consolidating range—chart courtesy of StockCharts.com

Managing Abnormal Probability

Fat tails—the visual outcome of abnormal distribution—are the reflection of how options trends and pricing cannot be distributed normally. With a fixed population and number of potential outcomes, statistical results include discrete random variables. Options trading, with its uncertain number of possible outcomes, is characterized by continuous random variables, which defines the unknown range of outcomes.

A solution to this problem is to define the *likely* range of outcomes based on historical volatility and price ranges over some period of time, Clearly, the less volatile the underlying price has been in recent history, the more reliable this assumption will be, when it allows for some movement outside of the established trading range. This probability density function is a device for managing the unknown degree of fat tails and limiting the analysis within what appears as a reasonable set of assumptions. For example, in a chart with a consolidating range, it is not difficult to limit the *likely* outcomes regarding future price levels. An example of this is shown in Fig. 9.2.

On the USO chart, the full scaling range resides between $12 and $9.20. Over 6 months, price moves between these high and low levels. However, a resistance zone between $11.20 and $11.80 is highlighted on the chart; and a support zone between $9.60 and $9.80 is also marked. Given these parameters, it is reasonable to also assume that future prices will trade within these

Fig. 9.3 Historical volatility, trending range

areas. The 50 MA line remained within the range for the entire period, and for most of the chart, prices only briefly moved above resistance or below support.

The limitation of the outcome range is more difficult in cases where the underlying is trending. For example, Fig. 9.3 provides an example of a chart with dynamic trends in place of a consolidating range.

The chart of Disney is scaled from a high of $107 to a low of $90. The first half of the chart is characterized by a descending channel, with the latter part an ascending channel. In order to limit the probability in this instance, it is necessary to anticipate a likely range based on price behavior and its average. The 50 MA tracked the descending channel precisely. However, as the price reversed and ascending, the 50 MA fell below. This can be used to identify a likely range taking recent price movement into account. Although this is a subjective approach, it appears reasonable to assume three points: (a) the support level is likely to remain at or above $92, the price at which the 50 MA departed from the ascending trend; (b) resistance level will level out as the 50 MA and price converge at some level, perhaps at or below $120 per share (calculated as a movement in price equal to the movement between reversal point of $92 and current price of $106); and (c) future price movement will be reasonably assumed to reside between $92 and $120 until new information is known (earnings changes, changes in divided, mergers or acquisitions, for example).

To identify the probability of price movement, all possible outcomes (in other words, any price from zero to infinity) represent the *sample space*. The actual underlying price on the closing of last trading day for an option is the outcome. In limiting the potential outcome, it becomes possible to develop a reasonable model of probability. As a first step, with equally likely outcomes, a probability formula can be greatly simplified[2]:

Probability with equally likely outcomes

$$E \div S = P$$

E number of outcome events
S number of outcomes in sample
P probability

This simplified model is a useful starting point for understanding options probability. However, knowing that a risk-neutral measure is not reliable nor realistic. This explains why options trading requires not a pricing model to articulate probability, but an identifiable probability limitation, or an expected value. It is necessary to limit the range of this expected value. For example, an underlying with assumed maximum trading range between $92 and $120 reflects a finite level of outcomes. Options with strikes within this range can then be articulated as well. Given this history of price ranges, it is *possible* that trading could fall to $3 per share or rise to $700. However, these are remote possibilities and may be excluded. The shorter the timeframe under review, the more remote these unlikely prices become. For example, a trader focusing on uncovered puts expiring in one to 2 weeks will be likely to focus on the immediate range and to not consider the higher or lower levels. Thus, the expected value of variables is restricted even though the theoretical price movement may be infinite.

Aspects of Probability

In previous chapters, various aspects of probability have been explored in several ways. The application of visual expressions of volatility, such as Bollinger Bands (BB), created a probability matrix. In Chap. 8, this concept was expanded by adding a second layer of BB to reflect three standard deviations. The comparison between two and three standard deviations defined a zone between the outer bands of each in the form of a narrow channel. In terms of probability, this area between the two outer bands may be defined as either a "danger zone" or an "opportunity zone," depending on whether calls or puts are at issue, and on whether positions are long and short. The probability function of this visual matrix is derived from the observations concerning price behavior. These include:

1. Price rarely moves above the two-standard deviation upper band or below the two-standard deviation lower band.
2. When price does move outside of these outer bands, it tends to retrace back into range, usually immediately but almost always within a few treading sessions.
3. When applying three standard deviations, price even more rarely moves above the upper band or below the lower band. If and when price does make a move into this area, retracement is usually imminent.

These observations of probability require little in the way of calculations, as their significance is visual and compelling. For the practical trader, interested in the timing of trades, the use of a probability matrix is effective for the timing of trades.

This technique is further supported by the definitive calculation of profit, breakeven and loss for each trade. When considering the maximum profit or loss, traders are able to apply value judgments to determine whether or not a particular trade offers the potential worth pursuing, and of equal importance, whether the risk level justifies the trade. The immediacy or remoteness of a profit zone or maximum profit is helpful in articulating risk, just as the same attributes of loss zone or maximum loss are essential in identifying a match between risk tolerance and risk itself.

This comparison between probability and risk is the step most often overlooked in trading. A trader is likely to calculate probability as a separate function or, if not performing a calculation, to intuitively judge probability based on the attributes of a trade. For example, the low probability of profit in a short-term long straddle and widely understood; and the high probability of exercise in a short straddle is equally apparent to an experienced trader. What may be less well known is a correlation between probability and risk.

While probability is either calculated in isolation or intuitively understood, the risk level itself may be considered separately or not at all. A trader who has determined that a particular strategy is "low risk" may fail to consider how that risk changes when probability evolves from one level to another. For example, a short put is considered relatively safe because it has the same market risk as a covered call. So a trader who does not want to invest in shares may consider the short put as an equally safe trade. However, certain risks, if overlooked, change the low risk of the uncovered put. If this position is entered shortly before an earnings announcement and if a large negative earnings surprise results, the uncovered put may turn out to be an exceptionally high-risk trade. Its realization occurs if that put moves deep in the money on the day of the earnings announcement.

In this example, a trader calculates probability in many ways. Among the issues considered are the following:

1. Uncovered puts have the same market risk as covered calls. As a result, the uncovered put is a low-risk trade.
2. The strike selected is many points out of the money. This buffer zone ensures that a move downward in the underlying will not move the put in the money.
3. The probability matrix found in Bollinger Bands increases confidence levels, with the put strike selected at or below the lower band.
4. Even if the underlying price moves in the money, the uncovered put can be closed or rolled without difficulty.

These calculations of probability overlooked related risks:

1. The uncovered call is low-risk even if the underlying price moves in the money. Ownership of the underlying covered the exercise risk in the call. The put is a low-risk trade only if the underlying price remains at or out of the money.
2. A strike selected many points out of the money provides a finite degree of safety. However, if the underlying price moves enough points, even a buffer zone is of limited value.
3. Selection of a strike below the lower band provides safety in most cases. However, this does not guarantee that the underlying will behave as expected.
4. An uncovered put can be rolled to avoid exercise, but it is not always possible to entirely offset a large ITM change in status. For example, if earnings are reported prior to expiration, a large negative surprise could create a loss in the short put.

When combining probability with risk, it is possible to avoid the unexpected. In the example of a short put, several steps could be taken to mitigate risk. These include modifying the short put to set up a calendar spreads or vertical spread; avoiding months in which earnings are reported; limiting the dollar amount at risk; and restricting positions to exceptionally short-term expiration terms (2 weeks or less, for example).

Human Nature and Risk

Beyond the mathematical calculation of risk (as expressed in profit, breakeven and loss models of strategies, for example), the element of human nature also has to be taken into account. Options traders are often puzzled by their own behavior, and the tendency to set well-articulated goals for entry and exit, only to violate those goals repeatedly. Why is this?

Human nature may set up perceptions related to risk and whether or not taking greater risks is a rational decision. This human nature effect is described in that "if conditions are perceived to be less risky, then people may take more risk, and if the conditions are perceived to be more risky, then the amount of risk taken may be reduced."[3]

Even with the most advanced calculations of probability, if risk is ignored then the exercise does not accomplish its intended purpose. Human nature contains numerous risk elements that should be folded into the timing decisions for entry and exit based on probability. Several aspects of the human element explain why traders exhibit certain patterns of behavior, often contrary to their well-defined goals.

For example, a fear appeal—usually describing a means to motivate people by warning them of a danger—is intended to instill fear in order to cause someone to take a preferred course of action. In trading, the most obvious fear appeal is that of losing money. If you do not properly assess risk, you will suffer a net loss, perhaps one you cannot afford. If you do not calculate probability, you will not know the degree of likelihood for profitable outcome.

The problem with fear of loss is that it often ceases to work as traders gain experience. As a result, risk awareness is reduced and replaced by a belief that experience replaces the need for caution. This occurs at all levels of expertise and is repeated at each stage. Even an experienced trader discovers, upon making an ill-timed trade, that losses remain possible even when all of the right steps have been taken. As part of every trader's "risk journey," the fear of loss is always present, or at least it should be. Without this, it becomes too easy to ignore risk tolerance boundaries. The process of identifying these boundaries is well described in a 2010 paper:

> A journey of a thousand miles, the old saying goes, starts with a single step. And even the most complex project or industry changing product begins with a single idea. But just as a traveler would be foolish to set out on a journey without packing provisions and studying a map to find the optimal route, a good idea isn't enough to make a successful business venture. Among the

many measures a company or organization should take before setting off on a new project, one of the most important is an assessment of the inherent risks. Risk is associated with all projects and business ventures taken by individuals and organizations regardless of their sizes, their natures and the time and place of execution and utilization.[4]

The fear appeal is essential in order to control risk levels. The calculation of probability is party of this risk journey, but by itself it is not enough. Any trader who moves forward based on partial information invites greater risks. The fear appeal is intended to be persuasive, based on recognition that fear is not pleasant. The appeal itself may be seen as a threat or as a warning of the potential danger in a course of action.[5]

The fear of loss, like all fears, tends to dissipate over time, as traders accumulate higher degrees of experience. However, in spite of the perception that experience reduces risk exposure, it merely replaces fear of loss with a different form of risk. Thus, *familiarity risk* sets up new dangers for experienced traders. Experience itself may blind the experienced trader to the real nature of risk in a particular action. For example, a trader who has executed numerous successful trades on the same underlying that the very real risks become invisible. This trader might fail to recognize that the trade executed many times is not as low-risk now as it was in the past. The fundamentals of the company might have changed, creating higher volatility and exposure to loss. Consider recent decades of change in fundamental value for many companies, such as Eastman Kodak, General Motors, or Sears. Each of these companies was held in the highest esteem in the past, but failed to maintain fundamental strength over time. A trader familiar with any listed company may easily ignore the dangers in changing fundamental status, so that a once safe trade was a "sure thing" in the trader's mind. However, today, that same trade might be very high-risk given the changes that have occurred.

A similar risk that speculators face is the appeal to risk-taking itself. This *desirable risk* tendency describes an attraction to risk, a common problem in options trading. A speculator prefers to trade in short duration sets and is less likely to want to hold positions in the underlying. Even though many options strategies involve hedged equity positions, part of the appeal to a speculator converts risks into desirable attributes of a trade. This may be experienced in seeking high-risk, high reward positions, focus on complex trades where simple trades might work just as well, or attraction to some technical indicators that demonstrate the trader's knowledge but might not add to the trader's profits. For example, a trader who fails to set an exit price for a long option ignores the logical progression of a trade. If the exit point is identified as a 50% gain (profit-taking) or a 50% loss (bail-out), the goal is rational. Whether or the trader is able to follow through and take the exit is a different issue. A trader attracted to risk as a desirable attribute might decide that tracking Fibonacci retracement, 50 MA, or implied volatility is less likely to be able to exit. Even if price behavior conforms to the goals (for example, if the price moves to the identified retracement level), it does not mean the trader will exit. In this case, desirable risk is the motivation, and realizing profits is secondary.

A trader who suffers from this tendency is also likely to classify "risk" individually. The risk of exercise, missed opportunity, or ever greater profits, or the risk in not having taken a position before a big move, become obsessions for some traders whose risk appetite cannot be satisfied. A speculator may be able to identify a specific risk in an options trade, but may also fail to weight one type of risk more urgently than another:

> It makes no sense to talk about "identifying risks." What we're really doing is defining risk sets and trying to estimate their properties. Changing from talking about "identifying" to talking about "defining" is a helpful reminder that we need to be precise about what each item is about …

> Those familiar ratings of "probability of occurrence" and "impact if it does occur" are illogical when applied to sets of risks where the impact of risks in the set is not equal—and they usually aren't. The probability rating might make sense, but the impact rating does not. It depends which risk or risks in the set have occurred. We need to use probability distributions of impact, or approximations of them.[6]

Some Risk Theories

Much study has been devoted to an attempt to define and understand risk. The protection motivation theory (PMT) is one example. It contains two segments, a threat appraisal and a coping appraisal. This combines an analytical view of risk severity and consequences of an outcome.[7]

PMT may be coordinated with attributes of fear appeals, defining:

> … the three critical components of a fear appeal to be (a) the magnitude of noxiousness of a depicted event; (b) the probability of that event's occurrence; and (c) the efficacy of a protective response. Each of these communication variables initiates corresponding cognitive appraisal processes that mediate attitude change.[8]

When applied to behavior among options traders, this set of risk observations is instructive. How is a threat perceived in a particular type of option trade? And if a loss occurs, how does a trader respond? For many, this process is over-simplified or even ignored. Traders may tend to not analyze the threat or consequences associated with risk, notably when a favorite type of trade is employed. A trader may come to believe that all necessary analysis was performed long ago, and that this risk universe is well understood (familiarity risk). As a consequence, when a risk does occur, the trader is surprised and simply does not know how to respond. A coping analysis after the loss involves a study of perceived benefits or costs in the form that response takes. For example, with an open short position, rolling forward may be an obvious choice, but a complete analysis of limited risks versus continued risk exposure might lead to a different conclusion: There are situations in which taking a loss and moving forward to a new trade is the most sensible course.

A related risk concept is called the general deterrence theory (GDT). In this theory, the understanding of expected costs or benefits is compared to the level of risk in behavior. A study of this tendency concluded that a fear of loss impacts behavior and, as a result, the deterrence itself may lead traders to avoid risky decisions.[9]

Just as every market has its own culture, the options market is defined in part by the attributes of risk that are involved in trading. Not everyone is well suited to the market, not merely due to levels of risk, but also due to the complexity of the options world and its attributes (calculations of outcomes, excessive theoretical pricing models, estimates such as that of implied volatility, the Greeks). All of these attributes are part of a cultural theory of risk. It is found in every market and in every organization. However, in the options world, some aspects of the culture define the cultural theory of risk that applies. This includes a focus on theory rather than on the practical, preference for complexity over simplicity, and attraction to complex strategies due not as much to potential profits as to the complexity itself.

This "love of complexity" in a sense makes membership in the options "club" exclusive because not every trader is willing or able to cope with it. Perception is that options are exceptionally high-risk due to the complexity of trading. In fact, the true risk is found in focus on complexity, when a more practical, less complex solution often clarifies the issues and reveals that risk covers a wide spectrum, based on the type of strategy selected, and whether options are used to speculate or to hedge.

In the cultural theory of risk, the most apparent attributes are based on belief rather than fact. The combination of conceptual framework and empirical studies are meant to explain risk and to reinforce the individual's understanding of risk levels and his or her place within the culture (in this case, the world of options trading). However, beyond the theoretical, the basic definition of risk is properly cast in decisions made with risk tolerance as the guide:

> There are various factors which affect the financial decision making of an individual of which demographic variables like age, gender and occupation and personal financial risk tolerance are the most important ones. Risk tolerance is a crucial factor that influences a wide range of financial decisions. Risk tolerance is defined as individuals willingness to engage in a financial activity whose outcome is uncertain.[10]

This definition of risk tolerance has to be expanded to also acknowledge three tendencies in behavioral finance. The first is termed the *heuristics*, an observation that decisions often are made illogically and based on intuition, rule of thumb, or approximation. The second attribute is *framing*, a cultural aspect of risk consisting of stereotypes or emotional filters resulting from common anecdotes, developed to help traders better understand events. In the markets, the well-known crown mentality is an example in which a majority acts in the same manner, often ill-timed and directionally wrong (for example, a tendency for traders to buy at the top or sell at the bottom, when a contrarian view is to do the opposite). The third tendency is *inefficiency*, which may be applied to mispricing of securities in a broad sense, or to irrational decisions on an individual level. Collectively, these three tendencies further define how traders respond to risk or act to perpetuate the behaviors (heuristics, framing or inefficiency).

Cultural theory is contrasted with rational choice theory, which recognizes perception of risk as a process sin which the individual weighs the offset between costs and benefits.[11] The theory is descriptive of outcomes rather than processes of choice. In balancing cost versus benefit of a choice, the goal is to achieve maximum advantage, what might also be termed an informed judgment call.[12]

The rational choice theory applies to options trading, notably when traders consider a particular trade in terms of risk. If this is combined with an analysis of probability, the comparison between costs and benefits is complete. It highlights the difference between a judgment call that is merely intuitive, versus a judgment call based on the combination of experience, knowledge, and probability testing.

Outcome analysis is complex in the sense that a broad range of possible outcomes exist for every options trade—degrees of profit or loss, the value of a hedge, or the role played by dividends, for example—but there is a tendency to see outcomes in a more binary manner:

> Outcomes are commonly perceived as positive or negative in relation to a reference outcome that is judged neutral. Variations of the reference point can therefore determine whether a given outcome is evaluated as a gain or as a loss. Because the value function is generally … steeper for losses than for gains, shifts of reference can change the value difference between outcomes and thereby reverse the preference order between options.[13]

The Role of Variance in Risk Perception

Without variance, there would be no risk, since variance is, by definition, the uncertainty inherent in any situation. In statistics, great effort is expended to quantify risk, and to add certainty to an uncertain set of facts. Charles Darwin's cousin, Francis Galton, attempted in 1883 to publish a rationale for the theory of eugenics. He studied the tendency of tall parents to have tall children, but only on average:

> Indeed, the mean height of the sons of exceptionally tall fathers tended to be slightly lower than the father's height – and closer to the population average – as if an inevitable force were always dragging extreme features toward the center. This discovery – called regression to the mean – would have a powerful effect on the science of measurement and the concept of variance.[14]

Galton is acknowledged as the theorist who discovered the statistical concept of regression toward the mean. This is the observed tendency for variance to move closer to the average of the entire field on each successive outcome.[15]

This statistical tendency is seen in options trading in many aspects. Any trend, for example, tends to level out as it nears completion and reversal. An option premium will interact with the underlying once its moneyness is closer, versus less responsive tendencies when deep in or out of the money. This also is a form of regression.

In options trading, "returns can be very unstable in the short run but very stable in the long run." Thus, the standard deviation of average returns tends to decline more rapidly than the inverse of a holding period. This is significant because it contradicts the random walk theory, the belief that all short-term price movement is impossible to predict. The observed reversion trend reveals, further, that when net returns have been low for a period of time, they are followed by offsetting periods of higher returns.[16]

As with other areas of study, the options trading version of regression to the mean simply observes that after a random event (i.e. a spike in option premium and also in the underlying) is likely to be followed by another random event that is not as extreme. For example, a trader who observes a strong increase in the underlying price and in option premium would be ill-advised to assume that another spike, equal in distance and in the same direction, will occur immediately after. It is far more likely that some or all of the price movement will return back (regress) toward the mean, or the previous price and premium levels. This is seen frequently after an earnings surprise. The underlying price reacts with a strong move, usually more than justified by the degree of surprise; and option premium follows. Later the same day or over a period of days, the stock and option prices reverse and move back toward the previously established range.

However, the tendency to revert does not prove that the opposite movement is a form of compensation for the spike in price. It is *likely* to occur, either entirely or to a degree. Options traders, like all market participants, are likely to fall victim to the *gambler's fallacy*, a belief that an unusual event will be offset by an opposite-moving event, or that a less frequent event is going to be followed by one of greater frequency. Those who track earnings surprises may convince themselves that a positive earnings surprise last quarter will lead to a negative surprise this quarter.

This assumption, also termed the Monte Carlo fallacy, is a common one among traders. Acting upon this false logic may lead a trader to ill-timed trades, or in making trades in the wrong direction (long versus short, for example), even when technical indicators provide evidence of greater reliability. This is one example of a mistaken assumption about reversion to the mean. It is not an attribute of karma in which events even out, but a long-term tendency closely associated with the law of large numbers. However, whereas the large numbers observation reports accurately on outcomes for a large population, it does not promise the same outcomes for individual events, such as stock price movement or related option premium reaction.

Variance—and the potential for relying on false logic—is an attribute of risk perception, and the danger of costly interpretation may be termed *decision risk*, the issues faced when several choices are available. Should you close a trade and take a minimal profit? Should you wait to see if higher profits accumulate? Are you willing to see profits turn to losses due to time decay? As a form of false logic, traders may choose to believe that a trade will somehow work out profitably, even as evidence accumulates that profit is the most unlikely. The decision risk has two parts: the risk of turning small profits into total losses, and the risk of missing opportunity if and when conditions improve *after* exiting the trade. This is a particular dilemma that recognizes risk but may easily overlook probabilities:

> Decision risk is a construct used to characterize the alternatives confronting a decision maker; it can, for example, describe how undesirable the likely effects of an alternative are and the likelihood of their occurrence. Risk can also be used to characterize an overall decision – how risky it is compared to other alternatives.[17]

Theory of Probability

The usual definition of *probability* is the likelihood that an event will occur. However, this definition should be expanded from its additive description to a multiplicative version: The likelihood of an event not occurring, but replaced with one of several known other outcomes; and the result then subtracted from '1' to determine the occurrence of the subject event.

At first glance, it would seem that these two definitions (an event occurring, or not occurring) would result in the same outcome. However, they do not. Repeating the simplified example from Chap. 1, the additive method of probabilities requires multiplying the number of attempts by the fraction of possible outcomes:

Additive probability

$$A * (1 \div x) = p$$

A attempts
x number of possible outcomes
P additive probability

For example, with four rolls of a single die, the outcome is:

$$4 * (1 \div 6) = 67\%$$

If the same formula is applied to 24 rolls of two dice, the outcome is the same:

$$24 * (1 \div 36) = 67\%$$

The adjustment to this formula was necessary, as Blaise Pascal discovered in the seventeenth Century, improves accuracy by applying the multiplicative method, in which the chances of the outcome *not* occurring are calculated, and the result subtracted from 1:

Multiplicative probability

$$1 - (O \div x)^n = P$$

O   negative outcomes
x   number of possible outcomes
n   number of attempts
P   additive probability

Applied to the example of four rolls of a single die, the chances of any one number coming up are calculated by using 5 as the number of negative outcomes, out of 6 possible outcomes, and 4 attempts:

$$1 - (5 \div 6)^4 = 52\%$$

The formula may also be applied using 24 rolls of two dice, in which case 35 is the number of negative outcomes, 36 is the number of total possible outcomes, and 24 is the number of attempts:

$$1 - (35 \div 36)^{24} = 49\%$$

The difference in additive and multiplicative outcomes is significant, 52 versus 67% for a single die rolled four times, and 49 versus 67% for two dice rolled 24 times.

This raises a question: How is probability calculated for options trading? Great effort is put into determining the probability of a particular option being in the money or out of the money on the last trading day, for example. This normally is calculated based on continuous probability distribution. In a discrete probability distribution, known numbers of outcomes are given weight and then added together. If, for example, there are six possible outcomes in a trial, and each has the same likelihood of occurring (for example, three rolls of a single die with 6 possible outcomes), the additive method reveals the discrete probability to be equal to the sum of each roll's outcome:

Discrete probability

$$A_1 + A_2 + \cdots A_n = P$$

A attempt
n number of attempts
P discrete probability

Applied to three rolls of a single die, the odds of any one number coming up are one in six:

$$1/6 + 1/6 + 1/6 = 50\%$$

The probability of any one number coming up is 50% in three throws of the die. The discrete probability is based on known positive or known outcomes. However, in continuous probability distribution, the possible outcomes are either not known or are infinite. With probability density function not specifically limited, the solution is to apply a cumulative distribution function which assumes limited values. As a result, probability resides within a range of values. This range, or interval, is then subjected to the calculation as the discrete probability. However, it is based on endpoint values of the identified intervals.; In other words, it is only an estimate. For example, the current underlying price is $50 per share. A call option is opened with a strike of 52.50. The probability of that option ending up in the money (with underlying at 52.51 or above) is compared with the chances of it being at the money (52.50) or out of the money (52.49 or below).

The flaw in this calculation of probability are immediately apparent. For continuous distribution, it is necessary to calculate the density function in place of the probability function. However, the method by which this is done may be subjective, depending on how the intervals are determined and set. To correctly assign probability, it is necessary to add appropriate weight to each outcome. What is the appropriate weight? Two points have to be concluded from the flawed calculation of probability that a particular option will be in the money on the last trading day: First, the calculation itself is based on an additive method, which is inaccurate; and second, the method of defining intervals and their weight is quite subjective, and may be arbitrary. A defense of the system used by most brokerage houses may be that the weighting is based on historical price volatility. However, every trader knows that historical prices are not cyclical. They are more random and caused by many unknown and known factors. A weather pattern, in comparison, is easily articulated by past year temperature intervals for a particular date. For arriving at the probability of where an option ends up (ITM, ATM or OTM), there is no science available, only estimation.

Interpreting Probability

The popular (and obvious) method for calculating probability is defined as quantified number of positive outcomes derived from a given number of events. As Blaise Pascal discovered, this is inaccurate and may mislead the interested person (trader, gambler, actuary, etc.) into making assumptions that are inaccurate.

Even beyond the easily proven error in additive probability calculations (versus the more accurate multiplicative system), two interpretations have to be taken into account. Once a calculation moves beyond the finite number of outcomes (coin toss or dice roll, for example), the characteristics of probability change as well.

The first interpretation recognizes that events are independent from one another with repeated attempts, but some finite quantity can be assigned to random events. Proponents of this view are called *objectivists* and the concept is based on recognizing the existence of truth outside of an individual's bias or interpretation. Objective probability looks for likely outcomes as relative frequency in a large number of attempts or outcomes.[18]

The second group is called *subjectivists*, probability can be quantified based on personal beliefs. This often is calculated based on a combination of expertise and experimentation. The theory assumes that, given enough data, all individuals eventually will arrive at similar conclusions. This is called *Cromwell's Rule*.[19]

This rule states that prior probabilities of '0' (an event will never occur) or '1' (the event will always occur) should be avoided, except when statements are logically true or false. For example, $3 + 3 = 5$ is logically false, whereas $3 + 3 = 6$ is logically true. When applied to probability in options trading, a subjective trader may contend that an underlying trading over the past year between $50 and $60 (and demonstrating no dynamic trend) is likely to trade between $50 and $60 in the near future—and that options with strikes within this price range contain subjective probability of not trading above or below that range. This is a personal belief and is used to set limits on the range of *likely* outcomes based on past price behavior, based on application of a *likelihood function* given the range a trader has defined.

The mathematical application of probability has to be based on a limited number of outcomes or, as in the case of options, an assumed *sample space* or range of likely outcomes. In applying a given sample space outcome, may have to be based on *conditional probability* or the probability that an event will occur based on the occurrence of another events. For example, consider the probability aspects of the risk of early exercise. The holder of an ITM short call during the week prior to ex-dividend date is aware that some portion of calls are exercised before ex-dividend as part of a dividend capture strategy. A long call holder may exercise against the short position and call away 100 shares. However, this does not always occur. Thus, there is a *conditional probability* of having a short call exercised, but it relies on two events: the call being in the money and the act of exercise.

One of several variations of conditional probability for problems such as this compare the intersection of two events, represented by the symbol ∩, with an assumption that event 'A' will occur given that event 'B' also occurs. This assumption is represented by the symbol | in the equation.

Conditional probability

$$P(A|B) = (P(A \cap B)) \div P(B)$$

P conditional probability
A event A
B event B
| assumption of both random events occurring
∩ the intersection of 'A' and 'B'

The entire study of probability is confounded by the many permutations of outcomes and by the variables that have to be based on great uncertainties. An option may end up in the money, but when and how far in the money? What time decay factors are in play based on timing, and even more uncertain, how will volatility affect premium levels? Given these variables, the possible outcomes (thus, the range of risk) is practically infinite. These uncertainties were observed in the expression, *Jedenfalls bin ich überzeugt, daß der Alte nicht würfelt* ("I am convinced that God does not play dice").[20]

Random Variables in Options Trading

The degree of random variables in the pure profit or loss analysis is considerable. However, beyond this, options traders face a less tangible form of random variables, having to do with processing and retrieval fluency. In options trading, many rules apply: those concerning relatives risk of long versus short, calls versus puts, long-term expiration versus short-term expiration, not to mention an array of dozens of strategies. Furthermore, depending on whether the purpose in trading is speculative or based on a desire to hedge equity portfolio risks, this *fluency variable* should not be overlooked. By definition, these attributes provide clues to how people behave:

> Processing fluency, or the subjective experience of ease with which people process information, is one such metacognitive cue that plays an important role in human judgment … Retrieval fluency is the subjective ease or difficulty with which people bring to mind exemplars that conform to a particular rule.[21]

Given the nature of options trading with its *infinite populations* of possible outcomes, continuous random variables Random variables exist of at please two dimensions. Infinite populations, which cannot be counted, are well-known in statistics. These also are found in options trading. Thus, the variables cannot be entirely understood until they are studied in the context of processing and retrieval fluency.

For example, a particular options trader may proceed with blind spots. Taking a position to *always* trade long options only and to *never* trade short options or to hold equity positions in the underlying, limits the possible range of profitable opportunities. This is an example of a poor processing fluency because the trader refuses to process information such as "short positions may also be possible in the right circumstances" or "some very conservative strategies require the combination of equity positions with short options."

As long as the trader is not willing to examine the full range of strategies and the potential for hedging, the only activity possible is long option speculation. This limits the probability universe as a consequence of attempting to limit the risk universe.

The same trader may suffer from retrieval fluency if it is impossible to acknowledge the exemplars that conform to a trading rule or to expand the point of view to at least consider whether some circumstances make short trading a less risky choice—a choice with a higher probability of success in some conditions.

Because options trading is an activity with infinite possible outcomes, adding the element of fluency within an assumed probability range only complicates the trader's ability to accurately estimate probability or to accurately define a range of risk. In other words, the ironic result of avoiding risk (selling short or owning the underlying) may lead to higher risk (long traders suffer from excessively expensive premium or rapid time decay). Whether the level of risk is more severe depends on circumstances, which is why a highly fluent trader is able to switch from one side to the other (long versus short) based on timing signals and technical indicators; or to adopt strategies based on those circumstances rather than a more isolated assumption of a strategy's inherent risk characteristics.

Chapter Summary

- skew and abnormal distribution characterize options probability
- fat tails visually display differences from normal distribution
- the human element and human behavior dramatically affect the risk universe
- risk theories have to be understood in terms of individual risk profiles
- variance directly affects a trader's risk perception
- fold the theory of processing and retrieval fluency into analysis of random variables Random variables.

Notes

1. Doane, David P. & Lori E. Seward. (2011) Measuring Skewness: A Forgotten Statistic? *Journal of Statistics Education* 19.2: 1–18.
2. Yates, Daniel S.; Daniel S. Moore, & Daren S. Starnes (2003). *The Practice of Statistics* (4th ed.) New York: Freeman. p. 308.
3. Parsons, K., A. McCormac. M. Butavicius & L. Ferguson (2010). Human factors and information security: Individual, culture and security environment. *Australia Government, Department of Defence. Command Control, Communications and Intelligence Division, Defense Science and Technology Organisation, Edinburgh, Australia.*
4. Ayyub, Bilal M., Peter G. Prassinos & John Etherton (2010). Risk-Informed Decision Making. *Mechanical Engineering, 132*(1), 28–33.
5. Maddux, James E. & Ronald W. Rogers (1983). Protection motivation and self-efficacy: A revised theory of fear appeals and attitude change. *Journal of Experimental Social Psychology.* 19 (5): 469–479.
6. Leitch, Matthew (2004). Rethink your attitude to risk—start to think about sets of risk. *Balance Sheet, 12*(5), 9–10.
7. Ifinedo, Princely (2012). Understanding information systems security policy compliance: an integration of the theory of planned behavior and the protection motivation theory. *Computers & Security*, 31:83–95.
8. Rogers, Ronald W. (1975). A protection motivation theory of fear appeals and attitude change. *The Journal of Psychology,* 91, 93–114.
9. Johnston, Allen C. & Merrill Warkentin (2010). Fear appeals and information security behaviors: an empirical study. *MIS Quarterly*, 34(3):549–566.

10. Chavali, Kavita & M. Prasanna Mohanraj, M. P. (2016). Impact of Demographic variables and Risk Tolerance on Investment Decisions-An Empirical Analysis. *International Journal of Economics and Financial Issues*, 6(1).
11. Starr, Chauncey (1969). Social benefit versus technological risk. *Science,* 165 (3899), 1232–1238.
12. Friedman, Milton. (1953) *Essays in Positive Economics.* Chicago: University of Chicago Press. pp. 15, 22.
13. Taversky, Amos & Daniel Kahneman (Jan. 30, 1981). The framing of decisions and the psychology of choice. *Science*, Vol. 211, Issue 4481, 453–458.
14. Mukherjee, Siddhartha (2016). *The Gene.* New York NY: Scribner. p. 68; and Galton, Francis (1886). Regression towards mediocrity in hereditary stature. *The Journal of the Anthropological Institute of Great Britain and Ireland.* 15: 246–263.
15. Stigler, Stephen M (1997). Regression toward the mean, historically considered. *Statistical Methods in Medical Research.* 6 (2): 103–114.
16. Siegel, Jeremy (2007). *Stocks for the Long Run, 4th Ed.* New York: McGraw-Hill, pp 13, 28–29.
17. Sitkin, Sim B. & Laurie R. Weingart (1995). Determinants of risky decision-making behavior: a test of the mediating role of risk perceptions and propensity. *Academy of Management Journal*, 38(6), 1573–1592.
18. Hacking, Ian (1965). *The Logic of Statistical Inference.* Cambridge UK: Cambridge University Press, pp. 36–37.
19. Jackman, Simon (2009). *Bayesian Analysis for the Social Sciences.* Hoboken NJ: John Wiley & Sons, p. 18.
20. Einstein, Albert (December 4, 1926) Letter to Max Born.
21. Alter, Adam L. & Daniel M. Oppenheimer (2009). Uniting the tribes of fluency to form a metacognitive nation. *Personality and Social Psychology Review.* 13 (3): 219–235.

10

Option Pricing Models

Chapter Objectives:

- evaluate the need for pricing models for options trading
- analyze the structure of the Black-Scholes (B-S) pricing model
- compare the original B-S model to modified versions
- consider the question of accuracy of B-S and its impact on trading decisions
- study the nine known flaws in B-S assumptions articulated by Fischer Black

Options theorists focus a great deal of time on *pricing models*, which are designed to estimate what the current value should be of a particular option contract. This chapter contains two hypotheses: First, pricing models are not needed; and second, the best-known of these, the Black-Scholes[1] (B-S) pricing model, contains so many flaws that cannot be used reliably.

The first question to address: "Do we need pricing models, and if so, why?"

No. Modeling is not needed, which is evident when the structure and nature of the market is studied. Options are derivatives, meaning their value is based on pricing and price movement of the underlying security:

> The value of equity options is derived from the value of their underlying securities, and the market price for options will rise or decline based on the related securities' performance. [2]

To track this hypothesis, begin with an analysis of the underlying security. How is it priced?

No pricing models are required to place value on underlying securities. The value of a share of stock, for example, is set by the open auction of the market. Stock prices are judged by both fundamental and technical indicators, including earnings, dividends, and the price/earnings ratio; prices move based on news, rumor, marketwide sentiment, and numerous other causes. However, there is no mathematical model to identify what a share of stock *should* be worth. Options, however, in spite of being derivatives of underlying prices, are assumed to require a pricing model.

The nature of the options market further defies the assumption that a price model is needed. Although options trading is at times described as part of an "auction market," prices are controlled and set by market markets in the NASDAQ and specialists in the NYSE. By definition, an auction market is based on supply and demand among buyers and sellers. For the stock market:

> An auction market is a market in which buyers enter competitive bids, and sellers enter competitive offers at the same time. The price at which a stock is traded represents the highest price that a buyer is willing to pay and the lowest price that a seller is willing to sell. Matching bids and offers are then paired together, and the orders are executed.[3]

The "making of a market" (thus, the name 'market maker') means the individual in that role enters bid and ask prices to offset disparities among buyers or sellers. Market makers buy or sell in their own accounts to smooth out the market, profiting from the bid/ask spread. In fact, the market makers set the bid/ask spread based on volume of trading. As a result, the level of profit market makers earn is based on how they set this range of pricing.

This raises a related aspect of option pricing. Many options traders rely on the mid-price between bid and ask to determine the current "value" of an option. Pricing models also are based on this fictitious price, even though the true price paid by a buyer or received by a seller will never be the mid-price. However, options pricing formulas invariably apply a single price for the option, without distinguishing between bid and ask.

Activity among buyers and sellers affects price level and movement within the bid/ask spread. As volume increases in trading, the spread diminishes as a factor of competition (among market makers) . However, the market does not function like a true auction market, where supply and demand act within the confines of *scarcity* and supply and demand trends. As long as the market maker serves as seller to every buyer, and as buyer to every seller, the resulting facilitated market and its bid/spread functions are controlled on the inside, and not led by supply and demand among the end buyer or seller. The forces of volatility in the underlying affect supply and demand and, as a result, also affect the market maker's bid/ask price setting.

In the stock market, an auction is constantly at work, with buyers and sellers interacting, and with profits for a broker derived from commission on each trade. In the options market, orders are placed with the broker, and the broker then buys from or sells to the market maker. In this format, it does not matter whether supply or demand is greater, because the market itself is *facilitated* by the market maker directly. This process is the opposite of an auction market, in which prices flow based on changes in supply and demand:

> Facilitation: The act of providing or preserving the liquidity of a market. For example, market makers facilitate markets when they trade on their own accounts to try to preserve an equilibrium of supply and demand.[4]

One description equates market makers with book makers who set odds for gamblers:

> … Market Makers are like the book makers in Las Vegas who set the odds and then accommodate individual gamblers who select which side of the bet they want. A Market Maker supplies a bid and ask price and then let *(sic)* the public decide whether to buy or sell at those prices. As an options trader, Market Makers are master position traders who aim to establish and profit from every low risk and risk free opportunities *(sic)*.[5]

The logical flow from the underlying security to the option reveals the great flaw in the primarily academic infatuation with pricing models. However, even if a pricing model were needed, the market's expectation would include a requirement that the model *accurately* describe an option's value. As the remainder of this chapter explains, the B-S pricing model includes not only one, but numerous flawed assumptions.

The Black-Scholes Pricing Model

The idea behind B-S and other pricing models is that options traders need a means for determining the fair price of an option. This starting point is flawed as an overall assumption. If options are derived from activity in the underlying security, and that security lacks a direct pricing model, then what is the basis for this need?

Fischer Black himself explained that the rationale for developing the formula was based on the desire to identify a perfect and riskless hedge:

> Suppose there is a formula that tells how the value of a call option depends on the price of the underlying stock, the volatility of the stock, the exercise price and maturity, and the interest rate ... Such a position will be close to riskless. For small moves in the stock in the short run, your losses on one side will be mostly offset by gains on the other side.[6]

This goal requires acceptance of many variables. The explanation by Fischer Black is understandable in the context of the underlying *theory* of pricing an option. Reality, however, brings all of the assumptions into question. In practice, the perfect, riskless hedge is not possible due to the unknown elements at play:

> *The total cost of hedging an option is known before hedging begins* ... [This] would be possible, however, if *all* the assumptions that go into the formula were correct. Theory, unfortunately, is a far cry from reality.[7]

The same message and conclusion has been explained in another way, comparing the value of the derivative to the current price of the stock, all aimed at anticipating the future movement in value of the option:

> The Black-Scholes model and others like it are theories that try to derive the *value* of an option so that it is consistent with the price of the underlying stock. They assume a market environment in which a dynamic riskless arbitrate strategy with the stock and the option is possible ... But in trying to apply a theoretical valuation model to the real world, it is immediately clear that none of the model assumptions actually holds.[8]

In fairness to Fischer Black, who with Myron Scholes developed the formula at the very time that options trading was first made available to the general public (and only for calls on a very limited number of stocks), developing pricing models for a new product and a new industry was pioneering in every respect. The effort was worthwhile as a theory, if only to demonstrate how a perfect hedge could be devised under ideal circumstances.

With this qualification in mind, the B-S formula is worth study if only to demonstrate the elusive nature of pricing, and the number of variables involved. However, since publication of the original paper, numerous adjustments have been made to expand the pricing model and to account for various considerations. These included calculating the pricing model for dividends, options on futures contracts, and options on currencies.

The formula for the Black-Scholes option pricing model is[9]:

$$c = SN(d_1) - Xe^{-rT}(d_2)$$

$$p = Xe^{-rT}N(-d_2) = SN(-d_1)$$

where

$$d_1 = (\ln(S \div X) + (r + \sigma^2 \div 2)T) \div \sigma\sqrt{T}$$

$$d_2 = [(\ln(S \div X) + (r - \sigma^2 \div 2)T) = d_1 - \sigma\sqrt{T}] \div \sigma\sqrt{T}$$

c	call (European style)
p	put (European style)
S	stock price
X	strike price of the option
r	risk-free interest rate
T	time to expiration (in years)
σ	volatility of the relative price change of the underlying stock price
$N(x)$	the cumulative normal distribution function.

Expansion of the Pricing Model

Although the pricing model has to be viewed as based on theory rather than for practical use, it serves a purpose. The model presents an idea about how a perfect hedge *could* work in the ideal situation—in which all of the variables conformed, the assumptions were correct, and the volatility of the underlying behaved exactly as expected.

The value in creation of a model is that it provides a starting point. Once the model has been developed, it can be used to set a standard. Just as a prospective home buyer knows the desired neighborhood, size and age of a home, and price, the search begins. Finding a house with all of the desired attributes is not always possible, but as a starting point, the modeling of an ideal provides guidance in the process of searching.

Since publication of the original B-S pricing model, several modifications were published to fine-tune the model. These modifications may be appreciated in the same context as the B-S: They provide further guidance in pursuit of a perfect hedge and a perfectly articulated price. Ignoring all of the variables (and the flaws in the underlying assumptions), these modified formulas continue the theoretical discussion of option pricing, but they do not add any value to the practical issues faced by options traders.

One modification was published in a 1973 paper written by Robert C. Merton. This adjusted formula includes a calculation of price for situations in which dividends are paid on a European option (dividends were not considered in the original formula).[10]

The 1973 Merton expansion of the Black-Scholes formula is:

$$c = S^{-qT}N(d_1) - X^{-rT}N(d_2)$$

$$p = Xe^{-rT}N(-d_2) - Se^{-qT}N(-d_1)$$

where

$$d_1 = \left[\left(\log(s \div x) + \left(r - q + \sigma^2 \div 2\right)\right)T\right] \div \sigma\sqrt{T}$$

$$d_2 = \left[\left(\log(s \div x) + \left(r - q - \sigma^2 \div 2\right)T\right) \div \sigma\sqrt{T}\right] = d_1 - \sigma\sqrt{T}$$

c	European call
p	European put
S	price of underlying asset
x	strike price
q	dividend yield
T	time to option expiration
N	cumulative normal distribution function
d	size of downward movement in the underlying
r	risk-free interest rate
σ	volatility in underlying price change

Another modification (Roll-Geske-Whaley) to B-S appeared in a series of papers published in 1977, 1979 and 1981. These resulted in a formula based on B-S for inclusion of dividend payments on American options.[11]

The Roll-Geske-Whaley formula is:

$$C = (S - De^{-rt})N(b_1) + (S - De^{-rt})M\left((a_1; -b_1;)\sqrt{t} \div T\right)$$

$$-Xe^{-rt}M\left((a_2; b_2;) - \sqrt{t} \div T\right) - (X - D)e^{-rt}N(b_2)$$

Where

$$a_1 = \left[\left(\ln\left(S - De^{-rt}\right) \div X\right) + \left(\left(r + \sigma^2\right) \div 2\right)T\right] \div \sigma\sqrt{T} \quad a_2 = a_1 - \sigma\sqrt{T}$$

$$b_1 = \left[\left(\ln\left(S - De^{-rt}\right) \div S\right) + \left(\left(r + \sigma^2\right) \div 2\right)T\right] \div \sigma\sqrt{T} \quad b_2 = b_1 - \sigma\sqrt{T}$$

$$c(S, X, T - t) = S + D - X$$

N	cumulative normal distribution
$M(a; b; \rho)$	cumulative bivariate normal distribution
S	asset price
X	strike price
D	dividend
t	time to dividend
T	time to expiration

A further modification to B-S was made in 1976 by Fischer Black to calculate a pricing model for options on forward and futures contracts.[12]

The 1976 Black forward and futures formula is:

$$c = e^{-rT}(FN(d_1) - XN(d_2))$$

$$p = e^{-rT}(XN(-d_2) - FN(-d_1))$$

where

$$d_1 = \left[\log(F \div X) + \left(\sigma^2 \div 2\right)T\right] \div \sigma\sqrt{T}$$

$$d_2 = \left[\log(F \div X) - \left(\sigma^2 \div 2\right)T\right] \div \sigma\sqrt{T} = d_1 - \sigma\sqrt{T}$$

c	price of the call
p	price of the put
e^{-rT}	discount factor
F	futures price
N	cumulative normal distribution
d	size of downward movement in the underlying
r	risk-free interest rate
σ	volatility in underlying price change

Another change was made to calculate pricing of currency options. In this formula devised in 1983 by Garman and Kohlhagen, the Merton-based dividend yield is replaced by a risk-free interest rate on a foreign currency, denoted as r_f.[13]

The Garman and Kohlhagen formula:

$$c = S^{-rf\,T} N(d_1) - X^{-rT} N(d_2)$$

$$p = Xe^{-rT} N(-d_2) - Se^{-rf\,T} N(-d_1)$$

where

$$d_1 = \left[\left(\log(s \div x) + \left(r - r_f + \sigma^2 \div 2\right)\right)T\right] \div \sigma\sqrt{T}$$

$$d_2 = \left[\left(\log(s \div x) + \left(r - r_f - \sigma^2 \div 2\right)T\right) \div \sigma\sqrt{T}\right] = d_1 - \sigma\sqrt{T}$$

c	European call
p	European put
S	price of underlying asset
x	strike price
r_f	risk-free interest rate on foreign currency
T	time to option expiration
N	cumulative normal distributionGarman and Kohlhagen formula function
d	size of downward movement in the underlying
σ	volatility in underlying price change

Many further modifications have been made over the years. Although these fine-tune the model and add to the literature of theoretical option pricing, none address the underlying problems, which are twofold. First is an alternative theory, introduced at the beginning of this chapter, that pricing models are not necessary when seeking answers beyond the theory. Second, the body of theories is inaccurate and does not reflect real-world outcomes.

Problems with Black-Scholes Inaccuracy

The value of a pricing model cannot be disputed. It is a valuable exercise when studying the theoretical model of options and how they *should* be priced, given a range of assumptions concerning volatility, risk-free interest, and more. However, practical application of the model does not yield guidance for traders seeking accurate pricing of options. For that, the options trader has to rely on the same forces affecting the underlying security: supply and demand, historical volatility, and stock valuation derived through fundamental and technical indicators.

The indicators relied upon by stock traders come in dozens of forms. One indicator that includes both fundamental and technical outcomes is the price/earnings ratio (P/E). This compares price (a technical signal) to earnings (the latest reported fundamental signal).

$$P \div E = P/E$$

P   price per share
E   latest reported earnings per share
P/E price/earnings ratio

The result of this formula is called the *multiple* because it represents the numbers of years of earnings reflected in the current price per share. For example, a P/E of 10 reveals that the current price per share is equal to 10 years of earnings per share, based on the latest report.

A "reasonable" P/E is normally defined as residing between 10 and 25. However, an accurate read has to take into consideration the problem with P/E. It compares the current price to the latest earnings, which may be several weeks out of date. The best way to judge P/E is by the long-term trend, studying the high and low P/E each year.

This example of how stock investors judge the current value per share as a bargain or as overpriced, can be carried over to judge options values as well. A well-priced underlying security is likely to have relatively well-priced options. This varies, of course, by emerging changes in historical volatility, moneyness of the option, and time to expiration.

With so many variables, a reliable method for spotting options values may be based on the combined study of price indicators (technical analysis) with the mathematical applications involving probability. These include the probability matrix visualized with Bollinger Bands, for example.

Why not just use the B-S price model or one of its variations?

The answer is that the pricing models are deeply flawed when traders attempt to spot bargain prices or overpriced conditions of options. For example, a rejection of B-S in a 2011 paper suggested that the model was not new, but merely set up a modification of existing pricing models to identify dynamic hedging as a means for eliminating risk.[14]

A strongly worded criticism B-S was offered in 2008 by Warren Buffett:

> The Black-Scholes formula has approached the status of holy writ in finance ... If the formula is applied to extended time periods, however, it can produce absurd results. In fairness, Black and Scholes almost certainly understood this point well. But their devoted followers may be ignoring whatever caveats the two men attached when they first unveiled the formula.[15]

The most noteworthy criticism of the B-S pricing model did not come from an outside observer, however. In an article published 15 years after the initial formula appeared, Fischer Black himself documented nine specific flaws in the formula. Black began his article by acknowledging that the model and actual price rarely match. He explained further:

> ... there are three reasons for a difference between value and price: we may have the correct value, and the option price may be out of line; we may have used the wrong outputs to the Black-Scholes formula; or the Black-Scholes formula may be wrong. Normally, all three reasons play a part in explaining a difference between value and price.[16]

As Black documented in his article, there are no fewer than nine flawed assumptions built into the model. A single flawed assumption makes a formula inaccurate, an obvious point. However, as additional flaws are added, the problem increases exponentially. With *nine* known flaws, the pricing model is simply unreliable for trading purposes. This further means that modifications intended to adjust the formula must also contain some of those same flaws. Although subsequent models have adjusted for many of these flaws, some of the more serious ones (such as an assumption of unchanging volatility) cannot be remedied.

These nine flaws are:

1. **Exercise**. Options are assumed to be exercised in the European method, meaning no early exercise is possible. In fact, U.S. listed stocks are exercised in the American model, meaning exercise may occur at any time prior to expiration. This makes the original calculation inaccurate.
2. **Dividends**. The underlying security is assumed to not pay a dividend. Today, many stocks pay dividends and, in fact, dividend yield is one of the major components of stock popularity and selection, and a feature affecting option pricing as well (This flaw in the original model was corrected after the initial publication once they realized that many stocks do pay dividends).

3. **Calls but not puts.** Modeling was based on analysis of call options values only. At the time of publication, no public trading in puts was available. Once puts began to trade, the formula was again modified. However, if traders continue relying on the original model, even for put valuation, they may be missing a fundamental inaccuracy in the price attributes.
4. **Taxes.** Tax consequences of trading options are ignored or non-existent. In fact, option profits are taxed at both federal and state levels and this affects net outcome directly. In some instances, holding the underlying over a 1-year period may lead to short-term capital gains taxation due to the nature of options activity, for example. The exclusion of tax rules makes the model applicable as a pre-tax pricing model, but that is not realistic. In fairness to the model, everyone pays different tax rates combining federal and state, that any model has to assume pre-tax outcomes.
5. **Transaction costs.** No transaction costs were applied to options trades. This is another feature affecting net value, since it's impossible to escape the brokerage fees for both entry and exit into any trade. This is a variable, of course; fee levels vary and, making it even more complex, the fee is reduced as the number of contracts traded rises. The model just ignored the entire question, but every trader knows that commissions can turn a marginally profitable trade into a net loss.
6. **Interest**. A single interest rate was applied to all transactions and borrowing; interest rates were assumed to be unchanging and constant over the life span of the option. The interest component of the model is troubling for both of these assumptions. Single interest rates do not apply to everyone, and the effective corresponding rates, risk-free or not, are changing continually. What might have been applicable, at least in theory, in 1973, is clearly not true today. Even adjusted pricing models since the original model tend to overlook this fact in how value is determined.
7. **Volatility.** The model assumes that volatility remains constant over the life span of an option. Volatility is independent of the price of the underlying security. This is among the most troubling of the model's assumptions. Volatility changes daily, and often significantly, during the option life span. It is not independent of the underlying and, in fact, implied volatility is related directly to historical volatility as a major component of its change. Furthermore, as expiration approaches, volatility collapse makes the broad assumption even more inaccurate.
8. **Trading is continuous**. Trading in the underlying security is assumed to be continuous and contains no price gaps. Every trader recognizes that price gaps are a fact of life and occur frequently between sessions. It would

be difficult to find a price chart that did not contain many common gaps. It is understandable that in order to make the pricing model work, this assumption was necessary as a starting point. But the unrealistic assumption further points out the flaws in the model.

9. **Price movement is normally distributed**. Price changes in the short term in the underlying security were assumed to be normally distributed. This statistical assumption is based on averages and the behavior of price; but studies demonstrate that the assumption is wrong. It is one version of the random walk theory, stating that all price movement is random. But influences like earnings surprises, merger rumors, and sector, economic and political news, all affect price in a very non-random manner. Underlying prices are simply not normally distributed.[17]

The topic of pricing models is controversial, but the evidence brings their value into question. Clearly, the flawed assumptions in B-S demonstrate that outcomes will not be reliable when applying the model. Furthermore, the question of whether a model is needed at all remains and brings up more questions.

For most traders employing options as portfolio hedges or for swing trading, the B-S model is not a practical tool for either selection of options or timing of trades. There are numerous reasons for this, including the tendency among swing traders to focus on soon-to-expire options. This means that volatility collapse is a serious inhibitor to any effort at determining option pricing. As expiration becomes short-term, the B-S pricing model becomes *more* inaccurate and tends to distort outcomes. For example:

> ... the Black Scholes option pricing formula exhibits pricing biases across moneyness and maturity. In particular, the Black Scholes formula underprices deep out-of-the-money puts and calls.[18]

This may further be the case even when volatility collapse is not an issue. However, it becomes more severe in the final weeks and days before expiration, rendering may swing trading strategies unreliable to the extent that the timing and selection of options is based on assumptions about pricing models.

Chapter Summary

- pricing models are not required for options trading
- the B-S model is best understood by studying its components
- several modifications to B-S have been published over many years

- the B-S model and its modifications contain known inaccuracies
- Fischer Black documented nine known flaws in the original formula.

Notes

1. Black, Fischer & Myron Scholes (May–Jun, 1973). The pricing of options and corporate liabilities *Journal of Political Economy*, Vol. 81, No. 3, pp. 637–654.
2. http://www.nasdaq.com/investing/options-guide/pricing-options.aspx.
3. www.investopedia.com.
4. http://financial-dictionary.thefreedictionary.com/Facilitation.
5. www.optiontradingpedia.com/market_makers.htm.
6. Black, Fischer (Winter, 1989). How we came up with the option formula. *Journal of Portfolio Management,* 15, 2.
7. Chriss, Neil A. (1997). *Black-Scholes and Beyond.* New York NY: McGraw-Hill. p. 120.
8. Figlewski, Stephen (Sep/Oct 1989). What does an option pricing model tell us about option prices? *Financial Analysts Journal,* 45(5), 12.
9. Black & Scholes, *Op. Cit.*
10. Merton, Robert C. (1973). Theory of rational option pricing. *Bell Journal of Economics and Management Science,* 4, 141–183.
11. Roll, Richard (1977). An analytic valuation formula for unprotected American call options on stocks with known dividends; Geske, Robert (1979). A note on the analytical formula for unprotected American call options on stocks with known dividends; and Whaley, Robert E. (1981). On the valuation of American call options on stocks with known dividends. *Journal of Financial Economics,* 5, 251–258; 7, 375–380; and 9, 207–211.
12. Black, Fischer. (1976) The pricing of commodity contracts. *Journal of Financial Economics,* 3, 167–179.
13. Garman, M. B. & S.W. Kohlhagen (1983). Foreign currency option values. *Journal of International Money and Finance,* 2, 231–237.
14. Haug, Espen Gaarder & Nassim Nicholas Taleb (2011). Options traders use (very) sophisticated heuristics, never the Black-Scholes-Morton formula. *Journal of Economic Behavior and Organization*, Vol. 77, No. 2.
15. Buffett, Warren, Letter to Shareholders, 2008, http://www.berkshirehathaway.com/letters/2008ltr.pdf.
16. Black, Fischer (March, 1988). The holes in Black-Scholes. *Risk.* 30–33.
17. Natenberg, Sheldon. (1994) *Option Volatility and Pricing.* New York NY: McGraw-Hill, pp. 400–401.
18. Henderson, Vicky, David Hobson, Sam Howison & Tino Kluge (2005). A comparison of option prices under different pricing measures in a stochastic volatility model with correlation. *Review of Derivatives Research,* 8(1), 5–2.

11

Alternatives to Pricing Models

Chapter Objectives

- review the history of the B-S pricing model and its development
- calculate simplified expected return to arrive at average outcomes
- master the underlying theory of CAPM as an alternative to pricing models
- compare simple and weighted moving averages to estimate volatility
- analyze the summary of options components
- conclude with a review of the 15 hypotheses of options trading.

A popular myth persists in the market that the Black-Scholes (B-S) pricing model was the first computation enabling traders to accurately price options. In fact, however, it was neither the first model, nor is it accurate.

In fact, many observers have disproven the validity of B-S on every level. These include a critique noted by widely-known experts in the financial markets. Referring to a 2007–2008 paper published by these experts (Espen Gaarder Haug and Nassim Nicholas Taleb) the article, published in *Fortune*, highlighted some key issues concerning the B-S model:

> The critique by Taleb and Haug ... contains three, innovative, status-quo-challenging dictums that do deserve wild popularization—if anything, as a way to get a widespread debate going. Pertaining to Black-Scholes, this is what these veteran players have to say:

- It's not used (even by those who think that they use it).
- It wasn't needed.
- It wasn't original in the first place.

Each of these bold assertions is conventional wisdom shattering. The three taken together in one devastating blow represent a potentially deadly strike to the most famous construct to ever emerge from financial economics ...[1]

The same article documented the true origins of options pricing formula and explained why B-S is not widely used by traders:

Black and Scholes (1973) and Merton (1973) actually never came up with a new option formula, but only a theoretical economic argument built on a new way of "deriving", rather rederiving, an already existing –and well known –formula. The argument, we will see, is extremely fragile to assumptions ... There are indeed two myths: (1) That we had to wait for the Black-Scholes-Merton options formula to trade the product, price options, and manage option books. In fact the introduction of the Black, Scholes and Merton argument increased our risks and set us back in risk management. More generally, it is a myth that traders rely on theories, even less a general equilibrium theory, to price options. (2) That we "use" the Black-Scholes-Merton options "pricing formula". We, simply don't.[2]

The article further explained that formulas for pricing options were not "invented" in 1973, but have been around for a long time:

The first identifiable [model] was Bachelier (1900). Sprenkle (1962) extended Bacheliers work ... James Boness (1964) ... derives a formula for the price of a call option that is actually identical to the Black-Scholes-Merton 1973 formula ...[3]

The disclosures by Haug and Taleb are profound, and they point why the flaws with assuming that B-S is the only answer to pricing, that it is even needed, and that it was the first instances of such a formula. However, it does not present alternatives. How does a trader know whether the current premium for an option is a bargain level or overly rich? Depending on whether the plan is to enter a long or a short trade (or a combination), knowing the richness of premium is essential as part of a timing strategy.

A pricing model is not the answer; but there are alternatives.

Alternatives to Pricing Models—CAPM

Given the known flaws and inaccuracies of B-S, many modifications and expansion have been published since 1973. However, expanding on a flawed model is questionable, versus seeking alternative methods for selecting options for trading.

Overall, an alternative is to focus on selection of the underlying based on fundamental indicators (defining profitability, capitalization and debt management, competitive strength, dividend trends, for example). Moving beyond the fundamentals, price volatility is defined through historical volatility and easily tracked. There is no need for implied volatility of an option when historical volatility of the underlying is readily available. Numerous studies and comparisons have concluded that the two versions of volatility result in similar outcomes; however, historical volatility is precise and measurable, concluding with a clear indication of the current level of volatility. In comparison, implied volatility (IV) is easily manipulated by changing the assumptions that go into the formula. It is an estimate only, and studies have concluded that IV adds little or no value to historical volatility: "… we found that implied volatility contains no more information regarding future realized volatility that is not already contained in recent historical volatility."[4]

An additional problem with IV is its tendency to exaggerate future volatility, based on the use of inaccurate assumptions:

> The forecast accuracy of implied volatility is inferior to the historical methods, as implied volatilities tend to overestimate future volatility. Regression results suggest that implied volatilities contain more information than historical volatilities, but that they are biased … Implied volatility does not impound all the information available from the past price history. The reverse tests of the orthogonality of historical volatility measures with implied volatility as an instrument indicate that call implied volatility would add little information to historical volatility …[5]

To overcome the problems of implied volatility as well as the overall inaccuracies in the B-S model, a reasonable alternative is found in the Capital Asset Pricing Model (CAPM).

CAPM compared the expected return on investments, with the risk involved. For options trading, the analysis and judgment concerning an options trade makes more sense bases on the risk/return analysis, than any attempt to identify the price through a flawed model. However, CAPM is not perfect any more than another model. It does allow traders to make a value judgment based on the crucial attributes of a particular option contract. If traders look only at expected return, or only at risk, they do not have a complete picture; and this is the value in CAPM.

Expected return is the assumed profit traders expect to earn from their investing and trading activity. This is a relatively simply exercise for a stock investment. Expected return may consist of the current dividend yield and assumed changes over time, added to expected stock price changes. For example, if the past three years have seen 0.5% increases in dividend yield consistently; and stock prices have risen by 4% per year over the same period, on average, it is rational to estimate expected return by continuing that three-year trend.[6]

A more complex expected return formulation adds together a range of assumptions, multiplied by potential percentages of outcome, and then calculating the average:

$$(P_1 + R_1) + \ldots (P_n + R_n) = E$$

where

$$R_1 + \ldots R_n = 100\%$$

P_1 potential outcome #1
P_n final potential outcome
R_1 realized result #1
R_n final result
E expected return

For an options trade, expected return is more elusive than the range of assumed possibilities for an investment in stock, given dividend yield and a *likely* range of results. An options trade has many more variables, traders have to assume the likely range, which may be based on assumptions, for example, of 100% profit, 50% profit, breakeven, 50% loss or 100% loss. If these outcomes are assigned varying potential outcomes expected return can be estimated:

100% profit at realized result estimated at 20%
50% profit at realized result of 30%
breakeven at realized result of 15%
50% loss at realized result of 10%
100% loss at realized result of 25%.

These assumed levels of realized profit may be random or based on a trader's actual experience. For example, the highest level of 30% for a 100% loss may be based on a history of not cutting losses prior to expiration; and the 50% profit is higher than the 100% profit because a trader has taken profits earlier than planned, rather than risking a reversal. To calculate expected return:

$$(100\% \times 20\%) + (50\% \times 25\%) + (0\% \times 15\%)$$
$$+ (-50\% \times 10\%) + (-100\% \times 30\%) = -0.025\%$$

Since this outcome is lower than zero, a trader would attempt to alter the outcome in a variety of methods:

1. Select contracts with higher than average net returns (increasing risk)
2. Write covered short positions on companies paying exceptionally high dividends (increasing overall return)
3. Adjusting goals regarding profit-taking and loss-cutting (reducing tendency to hold losing positions too long)
4. Evaluate selected strategies to determine whether alternatives would improve overall profitability (improving strategy selection).

All of these steps may help evaluate current trading policy to improve outcome. However, the analysis of profitability by itself through calculation of expected return does not take risk into account. By trading higher-volatility contracts, expected return might be improved, but at the same time the risk is also accelerated.

As long as the assumed outcomes are realistic, calculating expected return can be beneficial—if only to recognize a current flaw in strategy selection, entry or exit timing, or the use of one set of signals rather than another. For example, an analysis of outcomes over a defined past period and using those outcomes to calculate expected return in the future (assuming recent experience repeats) is one of several ways to use the expected return formula. However, by itself, the simplistic calculation fails to include other factors, such as a portfolio's standard deviation, beta, or time involved. This is where CAPM becomes useful in expanding expected return.

Combining expected return with risk rounds out the critical analysis of the trading program, viability of strategies, and individual goal-setting. The CAPM calculation combines expected return with evaluation of risk. The result offers a practical alternative to pricing models, which are designed to identify a riskless trade. However, just as the B-S model is based on many variables, CAPM also is limited by some of the same restrictions. B-S is based on the ideal of creating a riskless trade, but it does not address risk in a real sense.

The assumptions underlying CAPM contain many of the same flaws found in B-S. However, the one B-S flaw not associated with CAPM is the greatest of all, the assumption that volatility does not change over the life of an option.[7]

CAPM assumptions include: (1) there are no transaction costs; (2) assets are divisible, meaning the number of shares of stock or options is infinitely flexible; (3) no income taxes apply to returns; (4) the action of trading does not affect the price of the asset; (5) decisions are made solely on the basis of expected return and analysis of standard deviation in a portfolio; (6) there are no restrictions on going short; (7) lending or borrowing is possible at an assumed riskless interest rate; (8) traders are concerned with return variances in their trading activity; (9) all investors apply identical expectations regarding return on their portfolio; and (10) the asset at issue (options) are traded in a ready market and are easily marketable.[8]

The CAPM concept recognizes not only that risk has to be included in a formulation, but also the time value of money, expressed through the use of a risk-free interest rate. The Treynor-Mazuy Measure is one version of CAPM based on a combination of expected return, risk-free interest, and beta[9]:

$$(E(R_p) - R_f) \div \beta_p = T_p$$

$E(R_p)$ expected return
R_f return on a risk-free asset
β_p beta of the portfolio
T_p Treynor-Mazuy CAPM

This version of CAPM is more applicable to an equity portfolio than to an individual option trade; however, it may still be applied based on the assumed risk-free interest. The cautionary aspect of this method (as with any formulation) is that the risk-free rate is *assumed* to be an applicable equivalent return to the expected return of the option. Because this can be manipulated between different situations, it is one of those variables that may lead to confirmation bias and, consequently, a potential range of results not necessarily "accurate" for the purpose of the trade itself. No matter what method a trader uses to judge the risk and expected return, the variables are where the uncertainty resides. However, if applied consistently over many different possible trades, it yields a comparative outcome.

Rather than relying on beta as a means for calculating CAPM, an alternate method relies on standard deviation of returns. The Sharpe Measure was defined by its author as a reward-to-variability ratio.[10]

$$(E(R_p) - R_f) \div (\sigma(R_p)) = S_p$$

$E(R_p)$ expected return
R_f return on a risk-free asset
$\sigma(R_p)$ standard deviation of the portfolio returns
S_p Sharpe Measure CAPM

Any method employed to calculate the CAPM accepts the flaws of multiple assumptions. In this regard, CAPM contains similar flaws to the B-S model. However, it does not rely on the assumption of volatility (in B-S, this is assumed to be unchanged through the life of the option). Although CAPM is commonly applied to expected returns on equity portfolios, it can also be used to judge various options trading activity.

As an alternative to the B-S and similar models, CAPM can be applied to set up a consistent comparative method for critical evaluation. Because it does not rely directly on volatility (but rather on either beta or standard deviation), it is applicable to dissimilar underlying securities and options strategies. With B-S and its reliance on volatility assumptions, the reality that volatility is not the same for all securities makes comparison difficult, if not impossible. CAPM, in comparison, is consistent.

Difficulty of Volatility Forecasting

Volatility forecasting frustrates options traders, because at best all that is possible is a guess. It might be an informed guess, but the point remains that there are no accurate methods to anticipate future volatility. Implied volatility yields no better results than historical volatility, and yet the idea has been elevated to the status of certainty. It would be possible to make a forecast based on recent history of volatility; however, this should be undertaken with awareness of the fact that the result will only be an estimate.

For example, assume that volatility is studied for varying periods of time up until the present, and the following outcomes were identified in historical volatility:

Last 90 days 26%
Last 180 days 32%
Last 270 days 31%
Last 365 days 37%.

At first glance, it appears that volatility has been declining. However, with spikes due to earnings surprises or marketwide movement, is this reliable? Such events might be expected to recur in the near future. A simple moving average, in which a set number of values are taken into account and oldest dropped off as newer ones are used as replacements, could be applied to find one answer:

$$(V_{1+}V_{2+}\ldots V_f) \div N = A$$

V	values in the selected field
$1, 2, \ldots f$	first, second, remaining, and final values
N	number of values in the field
A	moving average

Applied to the previous example:

$$(26\% + 32\% + 31\% + 37\%) \div 4 = 31.5\%$$

However, in calculating averages, the latest entries often are assumed to earn greater weight than earlier values. Thus, a weighted moving average could be more accurate. This is calculated in several different ways. For example, the last entry could be given twice the value of previous entries:

$$(V_{1+}V_2 + \cdots (V_f * 2)) \div (N + 1) = A$$

V	values in the selected field
$1, 2, \ldots f$	first, second, remaining, and final values
N	number of values in the field
A	weighted moving average

An exponential moving average adds weight to the latest entry in a field of values through identification of an exponent and application to each subsequent value:

$$[((V_1 + V_2 \ldots V_f) \div N) - L] * (2 \div N) + ((V_1 + V_2 \ldots V_f) \div N) = NA$$

V	*values in the selected field*
1, 2, ... *f*	*first, second, remaining and final values*
N	*number of values in the field*
L	*latest entry*
NA	*new moving average*

Caution should be applied in attempting to estimate future volatility. In spite of the many efforts within the options industry to assure traders that volatility can be managed and anticipated, there is no actual certainty. The purpose of volatility estimation should be to enter comparative studies. Knowing that every underlying has a different volatility history, a consist method for identifying likely future volatility provides a degree of value, but never a certainty that the result is reliable.

For example, the previous discussion of weighting assumes that the latest volatility outcome is probably more accurate than older outcomes. This is not always a fair assumption. Statistical assumptions often are focused on the numerical valuation of a field of values, but not on the intangible factors that come into play. More than anything else, one of these factors is the time to expiration:

> short-term options, it may be true that data which covers short periods of time is *(sic)* the most important. But suppose we are interested in evaluating very long-term options. Over long periods of time the mean reverting characteristic of volatility is likely to reduce the importance of any short-term fluctuations in volatility.[11]

The averaging of volatility, whether weighted or not, is useful if only to confirm what many traders know intuitively. By understanding the nature of volatility and the rhythm and pattern displayed among a carefully tracked underlying, traders often have a strong understanding of the likely course of volatility, thus improving the timing of entry and exit based on volatility trends.

The fact that traders often apply intuition to anticipate volatility trends is an important aspect of effective trade timing. In theory, this could be captured in a formula; but the truth is that no formula can duplicate the experience-based understanding that traders have, in managing volatility effectively. On a practical level, traders are likely to undergo several steps to articulate volatility—and this is far more effective in appreciating the price of options, than any formula such as B-S could ever achieve. A current premium may represent a bargain price or be overly expensive, and these have opposite meaning to traders opening long trades versus those going short. The analysis of volatility can be broken down into several steps:

1. What has the underlying's historical volatility been over the long term?
2. How has recent historical volatility behaved in comparison to the average?
3. How do these outcomes affect the proposed option trade (short-term or long-term)?
4. Overall, has volatility appeared stable or erratic?

Volatility analysis may lead to strategic steps designed to exploit time decay, especially when time to expiration is very short. For example, in strategies combining a long and a short side, timing of each entry may be done in legs based on a tendency for time decay to occur at exact moments. Options tend to lose 34% of their remaining time value on the Friday one week prior to the last trading day, for example. Thus, a combination strategy (such as synthetic stock or vertical spreads) can be entered in two parts. The short side is opened on the Friday seven days before last trading day; and the long side opened on the next trading day, Monday. Both sides will lose 34% of their time value, on average, between these trading days, making the position more advantageous.

The risk of this approach is found in volatility itself. A highly volatile underlying might tend to move significantly between Friday's close and Monday's open, exposing the short side of the trade at the beginning of the last week of the option's life. However, a trader willing to take this risk will rationalize it due to the advantage of declining time value between Friday and Monday. This approach contains lower risk for underlying securities with relatively low volatility, which is one factor to consider in attempting to use this timing strategy.

This analysis of volatility and timing is more practical for timing of trades than any application of B-S to accomplish the same desired result. In fact, B-S does not articulate risk in the close-to-expiration scenario. It only attempts to identify a riskless price, but with so many variables at issue, this is not a practical concept.

Traders are most interested in determining how various options will perform if a trade is opened, based on a combined analysis of moneyness, expiration, and price behavior (volatility). To quantify this challenge, the basic attributes of volatility, probability, and the fundamental and technical bridge, represent a starting point in managing an options trading program.

Components of Basic Options Quantification (Chaps. 1–3)

A starting point in determining which option to trade, and its proximity (moneyness) is the great variable in options. Selection of strike, expiration and underlying present different levels of risk based on volatility.

Chapter 1 described the differences in types of volatility and made a case for preferring the certainty of historical, over the estimate of implied volatility. If a study of volatility is intended as a means for articulating probability of a profitable outcome, technical tools are available to help traders identify and compare this attribute. The *probability matrix* described in relation to Bollinger Bands (BB) aptly presents a visual summary of historical volatility. The two-standard deviation default of BB presents the matrix and enables traders to time both entry and exit. Expanding the default to overlay a three-SD version presents the extremes of price movement. This highly simplified probability matrix is an easy and obvious alternative to any attempt at defining what price should be based on estimated future (implied) volatility of the option. Since the option is derived from price behavior in the underlying, the option's probability is a function of underlying price volatility (historical). Using BB as a first step in determining the timing of trades supports the contention that the study of historical volatility (articulated in terms of option risks) is a rational method for selecting and timing trades, and for further understanding how degrees of risk are at play based on recent trends in historical volatility.

Chapter 2 expanded on the value of historical volatility by developing a correlation between the underlying price behavior and fundamental volatility (the tendency of key fundamental indicators to develop trends over time); and as a next step, a correlation between fundamental volatility and option risks. This concept—that fundamental trends and volatility directly affects option risks or creates opportunities—contradicts the widespread assumption that option risk is a separate and distinct matter, not connected to technical or fundamental features. However, the underlying price is based directly on both fundamental and technical trends and indicators but does not require a pricing model. Thus, a proximity rating system for fundamental and technical attributes of the underlying security present a method for timing of options trades based on the underlying. As a result, the derivative (option) may be expected to behave in a manner observable in the underlying (fundamentally and technically).

In Chap. 3, useful tools were added to the process, including put/call parity and its application in strategies such as synthetics; upper and lower bands and the effort to make the "risk universe" finite in options trading; Delta and Gamma as defining calculations; and the many variables associated with yield calculations.

Collectively, these three chapters lay the groundwork for a practical strategy, combining analysis with real-world concerns. This means trading with a focus on risk and probability and moving away from the theoretical (but popular) calculations such as implied volatility.

Applying the Basics to Manage Trades (Chaps. 4 and 5)

In the next two chapters, the difficulty of creating an effective system for selecting and managing trades was developed. Chapter 4 revealed the complexity of including dividend calculations to compare options among underlying securities. This total return approach too often is oversimplified, leading to inaccurate conclusions. The timing of quarterly dividends, comparative yields, and actual holding periods all affect combined option and dividend profitability.

Chapter 5 presented a realistic method for calculating the minimum an options trader needs to earn. The net after-inflation, after-tax breakeven is a reality check for anyone trading in securities and options. Many traders are not aware of the yield they need to beat inflation and taxes. This chapter also tackled the many calculations used to calculate return, and further examined methods for adjusting net breakeven price after rolling options forward.

These two chapters demonstrated how options traders are able to add accuracy to their analysis of returns, as well as calculations of risk. Added to the previous three chapters, this first half of the book provides a working foundation relying on realistic math and not on the theories that too often dominate the discussion of options trading.

Expanding the Basics to Define Profit and Loss Ranges (Chaps. 6–8)

In these chapters, a range of options strategies are described. In articulating any trade, knowing the levels of profit, breakeven and loss are essential starting points to define risks and opportunities, and to determine whether the structure of a trade is sensible. Only by understanding the risk elements of any trade can traders expect to succeed; and by matching the risk profile with individual risk tolerance, traders are more likely to select a range of trades appropriate for their circumstances, perceptions, and financial status.

Chapter 6 describes differences between speculation and hedging, and then applies the distinctions to a range of single-option trades. This most basic form of strategies is where most (if not all) traders begin. However, probability of profitable outcomes, even among long calls and puts, is not universally understood. Moving beyond the long call and put, the chapter further compares covered calls and uncovered puts; both strategies have the same market risk profile, and each may be used to hedge equity positions and to exploit price swings. However, covered calls are not immune from risk, as some traders mistakenly believe. Using covered calls and the closely related ratio write requires an appreciation of the risks involved.

Chapter 7 describes the many types of spreads used to speculate and to hedge equity positions. These include vertical, horizontal and diagonal spreads; condors and butterflies; and synthetic stock strategies. Each of these is described in terms of calculations for profit, breakeven and loss, and examples for each trade are further summarized in a payoff diagram.

Chapter 8 describes the same profit, breakeven and loss calculations for a range of straddles, and also concludes each with an example and a payoff diagram. Coverage includes long and short straddles, bullish and bearish combinations, covered straddles, strangles, strips and straps.

Conclusions of Risk and Pricing Models (Chaps. 9 and 10)

The preceding summary by chapter lays out a practical options strategy, based on math and focused on the needs of traders. With this in mind, the next two chapters compared the practical application of options trading with widely held misconceptions. Chapter 9 compared normal and abnormal distribution and demonstrated why statistical analysis (with finite fields of outcomes) cannot work with options analysis (and its infinite field of outcomes). Within the risk universe, the human element adds wisdom but may also interfere with objective analysis, so that this very aspect—the human element—is the most difficult variable of all. Processing and retrieval fluency are aspects to human nature, notably when dealing with random variables, that present great challenges to anyone who wants to create a world of certainty. With options trading, no models or mathematical formulas can lead to this result.

Chapter 10 presents the structure of the Black-Scholes (B-S) model and examines several of the modifications made to it over time. An unavoidable conclusion is that the model is not necessary, it does not work, and it is not widely used by options traders. This chapter also summaries the flaws in assumptions that the original authors overlooked in the original formula.

Conclusion

The premise for a practical approach to options trading can be summarized in a list of key hypotheses, which have been documented throughout this book. These are:

1. Implied volatility is not as accurate or applicable as historical volatility.
2. Methods used to calculate probability often are inaccurate, due to use of the additive method (chances of a single outcome occurring), in comparison to the more precise multiplicative method (calculation of outcomes *not* occurring).
3. Probability can be expressed visually as a *matrix* through standard deviation in indicators such as Bollinger Band.
4. Direct and clear correlations exist between fundamental volatility on the one hand, and both stock price and options behavior on the other hand. Thus, as a starting point, the study of fundamental trends assists traders in narrowing the search for profitable trades.
5. Some calculation, such as put/call parity, upper and lower bounds, Delta and Gamma, are useful in narrowing the field of options variables.
6. Dividend yield may be included in a calculation of options outcomes; however, this *total return* is not simple. The timing of ex-dividend dates with options expirations, makes annualized returns subject to timing and to holding periods.
7. Many traders do not calculate their net breakeven yield. This is often higher than many believe. A calculation emphasizes how options trading outperforms many other forms of trading, and assists in meeting and surpassing after-inflation and after-tax yield.
8. Options traders need to define whether their trades are designed to speculate or to hedge equity positions.
9. Basic single-option trades often are over-simplified, but analysis of the true risks as well as profit, breakeven and loss levels, will reduce risks and articulate appropriate trades based on risk tolerance.

10. Covered calls and uncovered puts contain identical market risks, but attributes of each trade present dissimilarities that traders need to understand.
11. Spreads, strangles and related strategies contain a broad range of risks and opportunities. By articulating the mathematical levels of profit, breakeven and loss, traders are equipped to better understand which strategies are most appropriate.
12. Options trading is statistically difficult to articulate, because outcomes are abnormally distributed. This reality demands clear definition of the human element in the risk universe.
13. Traders contend with fluency theory in defining their risk levels and seeking appropriate trades.
14. The Black-Scholes (B-S) pricing model is inaccurate and is not needed. Modifications to the original model cannot overcome the many flaws contained in the initial assumptions.
15. Alternatives to B-S exist. Rather than attempting to identify the "right price," these models present comparative pricing models. CAPM, for example, contains many of the sale flawed assumptions as B-S, but provides a more accurate system for comparing one possible expected return against another.

Chapter Summary

- the history of the B-S pricing model reveals its shortcomings
- simplified expected return helps develop estimates of average outcomes
- CAPM provides practical alternatives to pricing models
- simple and weighted moving averages can be used to estimate volatility
- options components as applied in preceding chapters set forth a useable program
- the 15 hypotheses of options trading are best summarized through mathematics.

Notes

1. Portema, Pablo Triana (April 8, 2008). Wither Black-Scholes? *Fortune*.
2. Haug, Espen Gaarder & Nassim Nicholas Taleb (November, 2007) Why We Have Never Used the Black-Scholes-Merton Option Pricing Formula, Third Version retrieved 1/23/17 @ http://www.mathsphi.com/article_Why_we_have_never_used_the_Black_Scholes_Merton_Option_Pricing_Formula_Haug_Taleb_nov_2007.pdf.

3. *Op. Cit*, referencing three sources: Bachelier, L. (1900). Theory of speculation in: Cootner, P. ed. (1964). The random character of stock market prices. Cambridge MA: MIT Press; Sprenkle, C. (1961). Warrant Prices as Indicators of Expectations and Preferences. *Yale Economics Essays*, 1(2), 178–231; and Boness, A. (1964). Elements of a Theory of Stock Option Value. *Journal of Political Economy*, 72, 163–175.
4. Kalyvas, Lampros & Nikolaos (2003). Causal relationship between FT-SE 100 stock index futures volatility and FT-SE 100 index options implied volatility. *Journal of Financial Management & Analysis, 16*(1), 20–26.
5. Gwilym, Owain Ap (2001). Forecasting volatility for options pricing for the U.K. stock market. *Journal of Financial Management & Analysis, 14*(2), 55–62.
6. Bernstein, Peter L. (May 1973). What rate of return can you 'reasonably' expect? *Journal of Finance*, 28, No. 2, 273–282.
7. Sharpe, William F. (Sept. 1964) Capital asset prices: A theory of market equilibrium under conditions of risk. *Journal of Finance, 19* (3), 425–442.
8. Fernandez, Pablo (Spring, 2015). CAPM: An Absurd Model. *Business Valuation Review*, Vol. 34, No. 1, pp. 4–23.
9. Treynor, Jack L. & Kay K. Mazuy (July–August, 1966). Can Mutual Funds Outguess the Market? *Harvard Business Review*, 131–136.
10. Sharpe, *Op. Cit.*
11. Natenberg, Sheldon (1994). *Option Volatility and Pricing*. New York: McGraw-Hill, pp. 279–280.

Appendix—Formulas

Additive probability

$$A^x(1 \div x) = p$$

- A attempts
- X number of possible outcomes
- P additive probability

Annual degree of change

$$(C-P) \div P = \%$$

- C current year
- P past year
- $\%$ percent of change

Annualized rate of return (option)

$$((C-O) - S) \div H * 365 = A$$

- C closing value
- O opening price
- S strike
- H holding period (days)
- A annualized rate of return

Annualized rate of return (stock)

$$((C-O) \div O) \div H * 365 = A$$

C closing value
O opening price
H holding period (days)
A annualized rate of return

Annualized return

$$R \div H * 365 = A$$

R return
H holding period (days)
A annualized return

Annualized return if exercised

$$(P \div S) \div D * 365 = A_O$$
$$((S_{100} - B) \div B)D * 365 = A_U$$
$$A_O + A_U = A_T$$

P premium of the option
S strike
D holding period (in days)
A_o annualized return if unchanged, option
A_U annualized return, underlying
A_T annualized return, total
B basis in the underlying for 100 shares
S_{100} strike for 100 shares

Annualized return if unchanged

$$(P \div S) \div D * 365 = A$$

P premium of the option
S strike
D holding period (in days)
A annualized return if unchanged

Annualized return (stock profits)

$$((E - B) \div B) \div H * 365 = A$$

E ending value
B beginning value
H holding period in days
A annualized return

Appendix—Formulas

At the money (ATM)

$$S = U = C_A$$
$$S = U = P_A$$

- U underlying price
- S option's strike
- C_A call ATM
- P_A put ATM

Bear call credit spread

Maximum profit:

$$P - F = M$$

- P premium received
- F trading fees
- M maximum profit

Breakeven:

$$S_s + M = B$$

- S_s strike, short call
- M maximum profit
- B breakeven

Maximum loss:

$$S_l - S_s - (P - F) = L$$

- S_l strike, long call
- S_s strike, short call
- P premium received
- F trading fees
- L maximum loss

Bear put debit spread

Maximum profit:

$$S_l - S_s - (C - F) = P$$

- S_l strike, long put
- S_s strike, short put
- C cost of the position
- F trading fees
- P maximum profit

Breakeven:

$$S_l - (C - F) = B$$

S_l strike, long put
C cost of the position
F trading fees
B breakeven

Maximum loss:

$$(C - F) = L$$

C cost of the position
F trading fees
L maximum loss

Black forward and futures formula

$$c = e^{-rT}(FN(d_1) - XN(d_2))$$
$$p = e^{-rT}(XN(-d_2) - FN(-d_1))$$

where
d_1 $[\log(F \div X) + (\sigma^2 \div 2)T] \div \sigma\sqrt{T}$
d_2 $[\log(F \div X) - (\sigma^2 \div 2)T] \div \sigma\sqrt{T}, \ = d_1 - \sigma\sqrt{T}$
c price of the call
p price of the put
e^{-rT} discount factor
F futures price
N cumulative normal distribution
d size of downward movement in the underlying
r risk-free interest rate
σ volatility in underlying price change

Black-Scholes pricing model

$$c = SN(d_1) - Xe^{-rT}(d_2)$$
$$p = Xe^{-rT}N(-d_2) = SN(-d_1)$$

where
d_1 $(\ln(S \div X) + (r + \sigma^2 \div 2)T) \div \sigma\sqrt{T}$
d_2 $[(\ln(S \div X) + (r - \sigma^2 \div 2)T), = d_1 - \sigma\sqrt{T}] \div \sigma\sqrt{T}$
c call (European style)

Appendix—Formulas

- *p* put (European style)
- *S* Stock price
- *X* Strike price of the option
- *r* Risk-free interest rate
- *T* Time to expiration (in years)
- σ Volatility of the relative price change of the underlying stock price
- *N(x)* The cumulative normal distribution function

Breakeven rate of return

$$I \div (100 - E) = B$$

- *I* rate of inflation
- *E* effective tax rate
- *B* breakeven rate of return

Bull calendar spread

Maximum profit:

$$U - S - (P + F) = M$$

- *U* underlying price by expiration
- *S* strike
- *P* premium
- *F* trading fees
- *M* maximum profit

Breakeven:

$$S - (P + F) = B$$

- *S* strike of the long call
- *P* debit paid for the long call
- *F* trading fees
- *B* breakeven, long call

Maximum loss:

$$P + F = M$$

- *P* debit paid for the position
- *F* trading fees
- *M* maximum loss

Bull call debit spread

Maximum profit:
$$(S_s - F) - (S_l + F) = P$$

S_s strike of short call
S_l strike of long call
F trading fees
P maximum profit

Breakeven:
$$S + D = B$$

S strike of the long call
D net debit paid
B breakeven

Maximum loss:
$$(P_l + F) - (P_s - F) = L$$

P_l premium, long option
P_s premium, short option
F trading fees
L maximum loss

Bull put credit spread
Maximum profit:
$$(S - F) - (B + F) = P$$

S sold put premium
F trading fees
B bought put premium
P maximum profit

Breakeven:
$$S - C = B$$

S strike of the short put
C net credit received
B breakeven

Maximum loss:
$$S_u - S_l - C = L$$

S_u strike (upper)
S_l strike (lower)
C net credit
L maximum loss

Cash flow to debt ratio

$$CE \div TD = CF$$

CE cash-based earnings
TD total debt
CF cash flow ratio

Conditional probability

$$P(A|B) = (P(A \cap B)) \div P(B)$$

P conditional probability
A event A
B event B
$|$ assumption of both random events occurring
$\cap$ the intersection of 'A' and 'B'

Condor:
Maximum profit

$$S_s - S_l - (D - F) = P$$

S_s strike, lower short call
S_l strike, lower long call
D debit paid
F trading fees
P maximum profit

Breakeven:

$$\text{Upper breakeven}: \quad S_u - (P - F) = B_u$$
$$\text{Lower breakeven}: \quad S_l + (P - F) = B_l$$

S_u Highest long strike
P Premium paid
F trading fees
B_u Upper breakeven
S_l Lowest long strike
B_l Lower breakeven

Maximum loss:
$$P + F = L$$

- P Premium paid
- F Trading fees
- L Maximum loss

Covered call:
Maximum profit:
$$P - F = M$$

- P premium
- F trading fees
- M maximum profit

Breakeven:
$$S - P = B$$

- S strike
- P net premium received
- B breakeven

Loss (decrease in underlying):
$$B - U = L$$

- B breakeven
- U underlying (below breakeven)
- L loss

Loss (increase in underlying):
$$U - P = B$$

- U underlying purchase price per share
- P net premium received for the call
- B net basis in stock

Covered put:
Profit:
$$B - C = P$$

- B basis in short stock (sold to open)
- C current price per share (bought to close)
- P profit

Maximum profit:
$$P - F = M$$

P premium of the short put
F trading fees
M maximum profit

Breakeven:
$$U + M = B$$

U underlying price at time of short sale
M maximum profit on the short put
B breakeven price

Loss:
$$U > B = L$$

U current underlying price
B breakeven price
L net loss

Covered straddle
Maximum profit:
$$(P - F) + S - U_b = M$$

P premium received
F trading fees
S strike
U_b basis in underlying
M maximum profit

Breakeven:
$$(U_b + S - (P - F)) \div 2 = B$$

U_b basis in underlying
S strike
P premium received
F trading fees
B breakeven

Maximum loss:
$$(U_b - U_c) + (P_c - P_b) = M$$

U_b underlying basis price
U_c underlying current price
P_c put, current premium
P_b put, net sale premium
M maximum loss

Debt capitalization ratio

$$D \div C = R$$

D long-term debt
C total capital
R debt capitalization ratio

Debt equity ratio

$$\text{Total liabilities} \div \text{Total equity}, = D/E$$

Debt ratio

$$\text{Total liabilities} \div \text{Total assets}, = DR$$

Delta

$$\Delta = \partial O \div \partial S$$

Δ Delta
∂O derivation of the option
∂S derivation of the stock

Delta relationship between call and put

$$\Delta_p = \Delta_c - 1$$

Δ Delta
p put
c call

Discrete probability

$$A_1 + A_2 + \cdots A_n = P$$

A attempt
n number of attempts
P discrete probability

Dividend yield

$$D \div P = Y$$

- D dividend per share
- P price per share
- Y dividend yield

Effective tax rate

$$T \div L = E$$

- T taxable income
- L tax liability
- E effective tax rate

Expected return

$$(P_1 + R_1) + \cdots (P_n + R_n) = E$$
$$\text{where} \quad R_1 + \cdots R_n = 100\%$$

- P_1 potential outcome # 1
- P_n final potential outcome
- R_1 realized result # 1
- R_n final result
- E expected return

Exponential moving average

$$[((V_1 + V_2 \cdots V_f) \div N) - L] * (2 \div N) + ((V_1 + V_2 \cdots V_f) \div N) = NA$$

- V values in the selected field
- $1, 2, \ldots f,$ first, second, remaining and final values
- N number of values in the field
- L latest entry
- NA new moving average

Gamma

$$\Gamma = \partial^2 V \div \partial^2 S$$

- Γ Gamma
- $\partial^2 V$ second derivation of the option
- $\partial^2 S$ second derivation of the underlying

Garman and Kohlhagen formula

$$c = S^{-rfT} N(d_1) - X^{-rT} N(d_2)$$
$$p = Xe^{-rT} N(-d_2) - Se^{-rfT} N(-d_1)$$

where

$$d_1 = \left[\left(\log(s \div x) + \left(r - r_f + \sigma^2 \div 2\right)\right)T\right] \div \sigma\sqrt{T}$$
$$d_2 = \left[\left(\log(s \div x) + \left(r - r_f - \sigma^2 \div 2\right)T\right) \div \sigma\sqrt{T}\right] = d_1 - \sigma\sqrt{T}$$

- c European call
- p European put
- S price of underlying asset
- x strike price
- r_f risk-free interest rate on foreign currency
- T time to option expiration
- N cumulative normal distribution function
- d size of downward movement in the underlying
- σ volatility in underlying price change

In the money (ITM)

$$U - S = C_I$$
$$S - U = P_I$$

- U underlying price
- S option's strike
- C_I call ITM
- P_I put ITM

Initial yield of an option

$$P \div S = R$$

- P premium
- S strike
- R return

Installment calendar spread
Required turns:

$$P_l \div P_s = N$$

- P_l premium, long option
- P_s premium, short option
- N number of turns required

Frequency required:

$$D \div T = F$$

D   days until LEAPS expiration
T   number of times required
F   frequency required

Interest coverage ratio

$$EBIT \div I = IC$$

EBIT   earnings before interest and taxes
I   annual interest expense
IC   interest coverage

Iron butterfly:
Maximum profit:

$$P - F = M$$

P   premium received
F   trading fees
M   maximum profit

Breakeven:

$$S_c + (P - F) = B_u$$
$$S_p - (P - F) = B_l$$

S_c strike of short call
S_p strike of short put
P   premium received
F   trading fees
B_u upper breakeven
B_l lower breakeven

Maximum loss:

$$S_l - S_s - (P - F) = M$$

S_l strike, long call
S_s strike, short call
P   premium received
F   trading costs
M   maximum loss

Iron condor
Maximum profit:

$$P - F = M$$

P   Premium received
F   Trading fees
L   Maximum profit

Breakeven:
$$S_c + P = B_u$$
$$S_p - P = B_l$$

S_c Strike, short call
S_p Strike, short put
P Net premium received
B_u Upper breakeven
B_l Lower breakeven

Maximum loss:
$$S_u - B_u = M \quad \text{or} \quad B_l - S_l = M$$

S_u strike, upper long call
S_l strike, lower long call
M maximum loss

Long butterfly
Maximum profit:
$$S_s - S_l - (P - F) = M$$

S_s strike, short call
S_l strike, lower long call
P premium paid
F trading fees
M maximum profit

Breakeven:
$$S_u - (P - F) = B_u$$
$$S_l + (P + F) = B_l$$

S_u strike, higher long call
S_l strike, lower long call
P premium paid
F trading fees
B_u upper breakeven
B_l lower breakeven

Maximum loss:
$$P - F = M$$

P premium paid
F trading fees
M maximum loss

Long call:
Profit:
$$U > B$$

U underlying price per share
B breakeven

Breakeven:
$$S + P = B$$

S strike
P premium paid
B breakeven price

Total loss:
$$P < S$$

P price of the underlying
S long call strike (net)

Long gut strangle:
Profit:
$$U - S_c - (P + F) = P_u$$
$$S_p - U - (P + FP) = P_l$$

U underlying price
S_c call strike
S_p put strike
P premium paid
F trading fees
P_u upper profit
P_l lower profit

Breakeven:
$$S_c + (P - F) = B_u$$
$$S_p - (P - F) = B_l$$

S_c call strike
S_p put strike
P premium received
F trading fees
B_u upper breakeven
B_l lower breakeven

Maximum loss:

$$(P + F) + (S_p - S_c) = M$$

P premium paid
F trading fees
S_p put strike
S_c call strike
M maximum loss

Long put:
Maximum profit:

$$S < (C + F) = P$$

S stock price
C premium paid
F trading fees
P profit

Profit:

$$(P + F) - C = N$$

P premium paid
F trading fees
C current price of underlying
N net profit

Breakeven:

$$S - (P + F) = B$$

S strike
P premium paid
F trading fees
B breakeven

Maximum loss:
$$P + F = M$$

P   premium of the long put
F   trading fees
M   maximum loss

Long straddle
Profit:
$$U - S - (P + F) = P_u$$
$$S - U - (P + F) = P_l$$

U   underlying price
S   strike
P   premium paid
F   trading fees
P_u upper profit (when $U > S - P + F$)
P_l lower profit (when $S < U - P + F$)

Breakeven:
$$S + P + F = B_u$$
$$S - P + F = B_l$$

S   strike
P   premium paid
F   trading fees
B_u upper breakeven
B_l lower breakeven

Maximum loss:
$$P + F = M$$

P   premium paid
F   trading fees
M   maximum loss

Long strangle
Profit:
$$U - S_c - (P + F) = P_u$$
$$S_p - U - (P + F) = P_l$$

Appendix—Formulas

U underlying price
S_c call strike
S_p put strike
P premium paid
F trading fees
P_u upper profit
P_l lower profit

Breakeven:
$$S_c + (P + F) = B_u$$
$$S_p - (P + F) = B_u$$

S_c call strike
S_p put strike
P premium paid
F trading fees
B_u upper breakeven
B_l lower breakeven

Maximum loss:
$$P + F = M$$

P premium paid
F trading fees
M maximum loss

Long strap
Maximum profit:
$$2 * (U - S) - (P + F) = M_u$$
$$S - U - (P + F) = M_l$$

U underlying
S strike
P premium paid
F trading fees
M_u maximum upper profit
M_l maximum lower profit

Breakeven:
$$S + (P + F) = B_u$$
$$S - (P + F) = B_l$$

S strike
P premium paid
F trading fees
B_u upper breakeven
B_l lower breakeven

Maximum loss:

$$P + F = M$$

P premium paid
F trading fees
M maximum loss

Long strip
Maximum profit:

$$U - S - (P + F) = M_u$$
$$(S * 2) - U - (P + F) = M_l$$

U underlying price
S strike
P premium paid
F trading fees
M_u maximum upper profit
M_l maximum lower profit

Breakeven:

$$S + (P + F) = B_u$$
$$S - (P + F) = B_l$$

S strike
P premium paid
F trading fees
B_u upper breakeven
B_l lower breakeven

Maximum loss:

$$P + F = M$$

P premium paid
F trading fees
M maximum loss

Appendix—Formulas

Lower bound (put)

$$P \leq S$$

P value of a put
S strike price of the option

Merton expansion of the Black-Scholes formula

$$c = S^{-qT} N(d_1) - X^{-rT} N(d_2)$$
$$p = Xe^{-rT} N(-d_2) - Se^{-qT} N(-d_1)$$

where

$$d_1 = [(\log(s \div x) + (r - q + \sigma^2 \div 2))T] \div \sigma\sqrt{T}$$
$$d_2 = [(\log(s \div x) + (r - q - \sigma^2 \div 2)T) \div \sigma\sqrt{T}] = d_1 - \sigma\sqrt{T}$$

c European call
p European put
S price of underlying asset
x strike price
q dividend yield
T time to option expiration
N cumulative normal distribution function
d size of downward movement in the underlying
r risk-free interest rate
σ volatility in underlying price change

Multiplicative probability

$$1 - (O \div x)^n = P$$

O negative outcomes
x number of possible outcomes
n number of attempts
P additive probability

Net basis in rolled option

$$S_c - (B_p - S_p) = A_c$$

S_c sell to open, new option
B_p buy to close, previous option

S_p sell to open, previous option
A_c adjusted basis, new option

Net basis in stock

$$B - P = N$$

B basis in stock
P premium received
N net basis

Net received or paid (option trade)

$$\text{Sell to open: } B - C = N$$
$$\text{Buy to close: } A + C = T$$

B bid price
A ask price
C trading costs
N net received
T total paid

Out of the money (OTM)

$$S - U = C_O$$
$$U - S = P_O$$

U underlying price
S option's strike
C_O call OTM
P_O put OTM

Payout ratio

$$D \div E = P$$

D dividend per share
E earnings per share
P payout ratio

Pearson's First Skewness Coefficient

$$(A - MO) \div SD = S$$

A mean (average)
MO mode

SD standard deviation
S   skewness coefficient

Pearson's Second Skewness Coefficient

$$(A - ME) \div SD = S$$

A   mean (average)
ME median
SD standard deviation
S   skewness coefficient

Percent of change (annual)

$$(C - P) \div P = \%$$

C   current year
P   past year
%   percent of change

Price bounds for calls and puts

$$Price\ bounds\ for\ a\ call = Z, U - E$$
$$Price\ bounds\ for\ a\ put = Z, E - U$$

Z   zero
U   underlying price
E   exercise price (strike)

Price/earnings ratio

$$P \div E = P/E$$

P   price per share
E   latest reported earnings per share
P/E price/earnings ratio

Probability with equally likely outcomes

$$E \div S = P$$

E   number of outcome events
S   number of outcomes in sample
P   probability

Put/call parity

$$C + X \div (1 + r)^t = P + U$$

C   premium of the call
X   strike
R   assumed interest rate
t   time to expiration
P   premium of the put
U   underlying price

Rate of return (option)

$$(C - O) \div S = R$$

C   closing value
O   opening price
S   strike
R   rate of return

Rate of return (stock)

$$(C - O) \div O = R$$

C   closing value
O   opening price
R   rate of return

Ratio write:
Maximum profit:

$$(n * (P - F)) = M$$

n   number of calls
P   premium of short calls
F   trading fees
M   maximum profit

Profit:

$$U \leq S = P$$

U   underlying
S   strike of the calls
P   profit

Breakeven:

$$S + M = B_u$$
$$S - M = B_l$$

S   strike of calls
M   maximum profit
B_u breakeven (upper)
B_l breakeven (lower)

Loss:
$$C - S = L$$

C   current price
S   strike price of uncovered portion
L   maximum uncovered call loss

Relative strength index (RSI)
$$100 - (100 \div (1 + RS))$$

(RS is the average of upward closings in the past 14 days, divided by the average of downward closings over the past 14 days.)

Return if exercised
$$(S - B + P) \div B = R$$

S   strike for 100 shares
B   basis in the underlying for 100 shares
P   premium of the covered call
R   return if exercised

Return if unchanged
$$P \div S = R$$

P   premium of the option
S   strike
R   return if unchanged

Risk freeze level
$$S - P = F$$

S   strike of the put
P   premium of the put
F   risk freeze level

Roll-Geske-Whaley formula

$$C = (S - De^{-rt})N(b_1) + (S - De^{-rt})M\big((a_1; -b_1;)\sqrt{t} \div T\big) \\ - Xe^{-rt}M\big((a_2; b_2;) - \sqrt{t} \div T\big) - (X - D)e^{-rt}N(b_2)$$

where

$$a_1 = [(\ln(S-De^{-rt}) \div X) + ((r+\sigma^2) \div 2)T] \div \sigma\sqrt{T} \quad a_2 = a_1 - \sigma\sqrt{T}$$
$$b_1 = [(\ln(S-De^{-rt}) \div S) + ((r+\sigma^2) \div 2)T] \div \sigma\sqrt{T} \quad b_2 = b_1 - \sigma\sqrt{T}$$

$$c(S, X, T-t) = S + D - X$$

N	cumulative normal distribution
$M(a; b; \rho)$	cumulative bivariate normal distribution
S	asset price
X	strike price
D	dividend
t	time to dividend
T	time to expiration

Sharpe Measure CAPM

$$\left(E(R_p) - R_f\right) \div \left(\sigma(R_p)\right) = S_p$$

$E(R_p)$	expected return
R_f	return on a risk-free asset
$\sigma(R_p)$	standard deviation of the portfolio returns
S_p	Sharpe Measure CAPM

Short butterfly
Maximum profit

$$P - F = M$$

P	premium received
F	trading fees
M	maximum profit

Breakeven:

$$S_u - (P-F) = B_u$$
$$S_l + (P-F) = B_l$$

S_u	strike, highest short call
S_l	strike, lowest short call
P	premium received
F	trading fees
B_u	upper breakeven
B_l	lower breakeven

Maximum loss:

$$S - S_l - (P-F) = M$$

- S strike of long calls
- S_l strike, lower strike short call
- P premium received
- F trading fees
- M maximum loss

Short gut strangle
Profit:

$$(P-F) + S_p - S_c = M$$

- P premium received
- F trading fees paid
- S_p put strike
- S_c call strike
- M maximum profit

Breakeven:

$$(P-F) + S_c = B_u$$
$$S_p - (P-F) = B_l$$

- P premium received
- F trading fees
- S_c call strike
- S_p put strike
- B_u upper breakeven
- B_l lower breakeven

Maximum loss:

$$U - S_c - (P-F) = M_u$$
$$S_p - U - (P-F) = M_l$$

- U underlying
- S_c call strike
- S_p put strike
- P premium received
- F trading fees
- M_u maximum upper loss
- M_l maximum lower loss

Short straddle
Maximum profit:
$$P-F = M$$

- P premium received
- F trading fees
- M maximum profit

Breakeven:
$$S + (P-F) = B_u$$
$$S - (P-F) = B_l$$

- S strike
- P premium received
- F trading fees
- B_u upper breakeven
- B_l lower breakeven

Maximum loss proximity:
$$\text{Upper loss occurs when: } U > S - (P-F)$$
$$\text{Lower loss occurs when: } U < S - (P-F)$$

- U underlying
- S strike
- P premium received
- F trading fees

Maximum loss:
$$U - S - (P-F) = M_u$$
$$S - U - (P-F) = M_l$$

- U underlying
- S strike
- P premium received
- F trading fees
- M_u maximum upper loss
- M_l maximum lower loss

Short strangle:
Maximum profit:
$$P-F = M$$

P premium received
F trading fees
M maximum profit

Breakeven:

$$S_c + (P-F) = B_u$$
$$S_p - (P-F) = B_l$$

S_c call strike
S_p put strike
P premium received
F trading fees
B_u upper breakeven
B_l lower breakeven

Maximum loss:

$$U - S_c - (P-F) = M_u$$
$$S_p - U - (P-F) = M_l$$

U underlying price
S_c call strike
S_p put strike
P premium received
F trading fees
M_u upper maximum loss
M_l lower maximum loss

Simple moving average

$$(V_1 + V_2 + \cdots V_f) \div N = A$$

V values in the selected field
1 2,… f, = first, second, remaining, and final values
N number of values in the field
A moving average

Standard deviation

$$\sigma = \sqrt{\frac{1}{N} \sum_{i=1}^{N} (x_i - \mu)^2}$$

σ standard deviation
N addition of values

Appendix—Formulas

Σ range of values from 1 to n
X_1 individual values
μ average

Synthetic long stock
Maximum profit:

$$U - S_c - (P + F) = M \,(\text{with net debit})$$
$$U - S_c + (P - F) = M \,(\text{with net credit})$$

U underlying price per share
S_c strike, long call
P premium paid
F trading fees
M maximum profit

Breakeven:

$$Sc + (P + F) = B \,(\text{with net debit})$$
$$Sc - (P - F) = B \,(\text{with net credit})$$

S_c strike, long call
P premium paid
F trading fees
B breakeven

Maximum loss:

$$Sp - U + (P - F) = M \,(\text{with net debit})$$
$$Sp - U - (P - F) = M \,(\text{with net credit})$$

Sp strike, short put
U underlying price
P premium paid
F trading fees
M net loss

Synthetic short stock
Maximum profit:

$$Sp - U - (P + F) = M \,(\text{with net debit})$$
$$Sp - U + (P - F) = M \,(\text{with net credit})$$

Sp strike, long put
U underlying price per share

P   premium paid
F   trading fees
M   maximum profit

Breakeven:
$$S_p - (P - F) = B \text{(with net debit)}$$
$$S_p + (P + F) = B \text{(with net credit)}$$

S_p strike, long put
P   premium paid
F   trading fees
B   breakeven

Maximum loss:
$$U - S_c + (P + F) = L \text{(with net debit)}$$
$$U - S_c - (P - F) = L \text{(with net credit)}$$

U   underlying
S_c strike, short call
P   premium
F   trading fees
L   loss

Taxable income
$$I - A - D - E = T$$

I   total income
A   adjustments
D   deductions
E   exemptions
T   taxable income

Time value
$$P - I - V = T$$

P   total premium
I   intrinsic value
V   implied volatility
T   time value

Treynor-Mazuy CAPM
$$\left(E(R_p) - R_f\right) \div \beta_p = T_p$$

$E(R_p)$ expected return
R_f return on a risk-free asset
β_p beta of the portfolio
T_p Traynor-Mazuy CAPM

Uncovered call:
Maximum profit:

$$P - F = M$$

P premium
F trading fees
M maximum profit

Profit:

$$U \leq S$$

U underlying value
S strike

Breakeven:

$$S + P = B$$

S strike
P net premium received
B breakeven

Maximum loss:

$$U - S - P = L$$

U underlying price
S strike price
P net premium received
L net loss

Uncovered call loss (if exercised)

$$C - S - P = L$$

C current price at the time of assignment
S strike of the uncovered call
P premium received for the call
L uncovered call loss

Uncovered put:
Maximum profit:
$$P - F = M$$

- P premium
- F trading fees
- M maximum profit

Breakeven:
$$S - P = B$$

- S strike
- P net premium received
- B breakeven

Loss:
$$B - U = L$$

- B breakeven
- U underlying price
- L partial loss

Upper bound (call)
$$C \leq S$$

- C value of a call
- S current price of stock

Variable ratio write:

Maximum profit:
$$P_{ls} + P_{hs} - F = M$$

- P_{ls} premium, lower strike
- P_{hs} premium, higher strike
- F trading fees
- M maximum profit

Profit:
$$U \leq S_l = P$$

- U underlying
- S_l lower strike
- P profit

Breakeven:

$$S_u + M = B_u$$
$$S_l - M = B_l$$

S_u higher strike calls
S_l lower strike calls
M maximum profit
B_u breakeven (upper)
B_l breakeven (lower)

Loss range:

$$\text{Upper price loss}: U > B_u$$
$$\text{Lower price loss}: U < B_l$$

U underlying
B_u breakeven (upper)
B_l breakeven (lower)

Loss:

$$\text{Upper price loss}: C - B_u = L_u$$
$$\text{Lower price loss}: B_l - C = L_l$$

C Current price
B_u breakeven (upper)
B_l breakeven (lower)
L_u loss (upper)
L_l loss (lower)

Weighted moving average

$$\left(V_1 + V_2 + \cdots \left(V_f * 2\right)\right) \div (N+1) = A$$

V values in the selected field
1 $2,\ldots f,$ = first, second, remaining, and final values
N number of values in the field
A weighted moving average

Bibliography

Aggarwal, N., & Gupta, M. (2013). Portfolio hedging through options: Covered call versus protective put. *Journal of Management Research, 13*(2), 118.

Alter, A. L., & Oppenheimer, D. M. (2009). Uniting the tribes of fluency to form a metacognitive nation. *Personality and Social Psychology Review, 13* (3).

Anson, M. J. P., Fabozzi, F. J., & Jones, F. J. (2010). *The handbook of traditional and alternative investment vehicles: Investment characteristics and strategies.* New York: Wiley.

Augen, J. (2009). *Trading options at expiration.* Upper Saddle River NJ: FT Press.

Ayyub, B. M., Prassinos, P. G., & Etherton, J. (2010). Risk-informed decision making. *Mechanical Engineering, 132*(1).

Bachelier, L. (1900). Theory of speculation. In P. Cootner (Ed.) (1964). *The random character of stock market prices.* Cambridge MA: MIT Press.

Bernstein, P. L. (1973, May). What rate of return can you 'reasonably' expect? *Journal of Finance, 28*(2), 273–282.

Black, F. (1975). Fact and fantasy in the use of options. *Financial Analysts Journal, 31*(4).

———. (1976). The pricing of commodity contracts. *Journal of Financial Economics, 3*, 167–179.

———. (1988, March). The holes in Black-Scholes. *Risk*, 30–33.

———. (1989, Winter). How we came up with the option formula. *Journal of Portfolio Management, 15*, 2.

Black, F., & Myron, S. (1973, May–Jun). The pricing of options and corporate liabilities. *Journal of Political Economy, 81*(3), 637–654.

———. (1974, May). The effects of dividend yield and dividend policy on common stock prices and returns. *Journal of Financial Economics, 1*(1).

Bollinger, J. (2001). *Bollinger on Bollinger bands.* New York: McGraw-Hill.

Boness, A. (1964). Elements of a theory of stock option value. *Journal of Political Economy, 72,* 163–175.

Bouchard, J.-P., & Potters, M. (2009). *Theory of financial risk and derivative pricing: From statistical physics to risk management* (2nd ed.). Cambridge UK: Cambridge University Press.

Buffett, W., Letter to Shareholders. (2008). http://www.berkshirehathaway.com/letters/2008ltr.pdf.

Calio, V. (2005). Reg FD eases effects of earnings sticker shock. *Pensions & Investments, 33*(7).

Canina, L., & Stephen, F. (1993). The informational content of implied volatility. *The Review of Financial Studies, 6*(3).

Chaput, J., & Ederington, L. H. (2003, Summer). Option spread and combination trading. *Journal of Derivatives, 10*(4).

Chavali, K., & Mohanraj, M. P. (2016). Impact of demographic variables and risk tolerance on investment decisions-an empirical analysis. *International Journal of Economics and Financial Issues, 6*(1).

Chicago Board Options Exchange (CBOE). http://www.cboe.com/learncenter/concepts/beyond/expiration.aspx.

Chriss, N. A. (1997). *Black-Scholes and beyond.* New York: McGraw-Hill.

Conrad, J. (1989, June). The price effect of option introduction. *Journal of Finance, 44.*

Coval, H. D., & Tyler, S. (2000, June). Expected options returns. *University of Michigan Business School.*

Daly, K. (2011, October). An overview of the determinants of financial volatility: An explanation of measuring techniques. *Modern Applied Science, 5*(5).

Derman, E. (2007). Sophisticated vulgarity. *Risk, 20*(7).

Doane, D. P., & Seward, L. E. (2011). Measuring skewness: A forgotten statistic? *Journal of Statistics Education, 19* (2).

Doran, J. S. (2007). The influence of tracking error on volatility risk premium estimation. *The Journal of Risk, 9*(3).

Einstein, A. (1926, December 4). Letter to Max Born.

Elenbaas, T., & David, T. (2006, Fall). Risk management for option writers. *Futures, 35.*

Enke, D., & Amornwattana, S. (2008). A hybrid derivative trading system based on volatility and return forecasting. *The Engineering Economist, 53*(3).

Eves, H. (1990). *An introduction to the history of Mathematics.* Fort Worth TX: Saunders College Publishing.

Farlex Financial Dictionary. (2009). Retrieved December 13 2016 from http://financial-dictionary.thefreedictionary.com/double+option.

Fernandez, Pablo. (2015, Spring). CAPM: An Absurd model. *Business Valuation Review, 34*(1), 4–23.

Figlewski, Stephen. (1989, September/October). What does an option pricing model tell us about option prices? *Financial Analysts Journal, 45*(5), 12.

———. (2004). *Forecasting volatility.* New York: University Stern School of Business.

Fisher, G. S. (2009, October 14). How to protect investments from cataclysmic 'fat tails'. *Forbes*.

Frederick, R. (2007, March). Trading with ratio spreads. *Futures*.

Freeman, M., & Freeman, K. (1993). Considering the time value of money in breakeven analysis. *Management Accounting, 71*(1).

Friedman, Milton. (1953). *Essays in positive economics*. Chicago: University of Chicago Press.

Galton, F. (1886). Regression towards mediocrity in hereditary stature. *The Journal of the Anthropological Institute of Great Britain and Ireland, 15*.

Garman, M. B., & Kohlhagen, S.W. (1983). Foreign currency option values. *Journal of International Money and Finance, 2*, 231–237.

Geske, R. (1979). A note on the analytical formula for unprotected American call options on stocks with known dividends. *Journal of Financial Economics*, 375–380.

Glazer, J. (1994, April). The strategic effects of long-term debt in imperfect competition. *Journal of Economic Theory, 2*(2).

Goltz, F., & Lai, W. N. (2009). Empirical properties of straddle returns. *Journal of Derivatives, 17*(1).

Goodman, T., Neamtiu, M., & Zhang, X. F. (2012, September). Fundamental analysis and option returns. *The Hong Kong University of Science and Technology*.

Goyal, A., & Saretto, A. (2013, April). Cross-section of option returns and volatility. *Journal of Financial Economics, 108*(1).

Grimes, Adam. (2012). *The art & science of technical analysis: Market structure, price action and trading strategies*. Hoboken NJ: John Wiley & Sons.

Grytam, T., Ng, S., & Francis, T. (2016, August 4). Companies routinely steer analysts to deliver earnings surprises. *The Wall Street Journal*.

Guay, W., & Harford, J. (1998, December). The cash-flow permanence and information content of dividend increases versus repurchases. *Journal of Financial Economics, 57*.

Gwilym, O. Ap. (2001). Forecasting volatility for options pricing for the U.K. stock market. *Journal of Financial Management & Analysis, 14*(2), 55–62.

Hacking, Ian. (1965). *The logic of statistical inference*. Cambridge, UK: Cambridge University Press.

Han, K. C., CFP., & Heinemann, A. (2008). A bull call spread as a strategy for small investors? *Journal of Personal Finance, 6*(2).

Haug, E. G., & Taleb, N. N. (2007, November). Why we have never used the Black-Scholes-Merton option pricing formula, Third Version retrieved 1/23/17 @ http://www.mathsphi.com/article_Why_we_have_never_used_the_Black_Scholes_Merton_Option_Pricing_Formula_Haug_Taleb_nov_2007.pdf.

———. (2011). Options traders use (very) sophisticated heuristics, never the Black-Scholes-Morton formula. *Journal of Economic Behavior and Organization, 77*(2).

Henderson, V., Hobson, D., Howison, S., & Kluge, T. (2005). A comparison of option prices under different pricing measures in a stochastic volatility model with correlation. *Review of Derivatives Research, 8*(1), 5–2.

Hirshleifer, J. (1977, September). The theory of speculation under alternative regimes of markets. *The Journal of Finance, 32*.

Hull, J. C. (2012). *Options, futures and other derivatives* (8th ed.). New York: Prentice Hall.

Hwang, S., & Satchell, S. E. (2000). Market risk and the concept of fundamental volatility: Measuring volatility across asset and derivative markets and testing for the impact of derivatives markets on financial markets. *Journal of Banking and Finance, 24*, 759–785.

Ifinedo, P. (2012). Understanding information systems security policy compliance: an integration of the theory of planned behavior and the protection motivation theory. *Computers & Security, 31*.

Israelov, R., & Nielsen, L. N. (2014). Covered call strategies: One fact and eight myths. *Financial Analysts Journal, 70*(6).

Jackman, S. (2009). *Bayesian analysis for the social sciences*. Hoboken, NJ: Wiley.

Johnston, A. C., & Warkentin, M. (2010). Fear appeals and information security behaviors: an empirical study. *MIS Quarterly, 34*(3).

Jorion, P. (1995, June). Predicting volatility in the foreign exchange market. *The Journal of Finance, 50*(2).

Kalyvas, L., & Dritsakis, N. (2003). Causal relationship between FT-SE 100 stock index futures volatility and FT-SE 100 index options implied volatility. *Journal of Financial Management & Analysis, 16*(1), 20–26.

Klement, J. (2011). Investment management is risk management-nothing more, nothing less. *The Journal of Wealth Management, 14*(3).

Kudryavtsev, A., Eval, L., & Shosh, S. (2014). Effect of inflation on nominal and real stock returns: A behavioral view. *Journal of Advanced Studies in Finance, 5*(1).

Leitch, M. (2004). Rethink your attitude to risk—start to think about sets of risk. *Balance Sheet, 12*(5).

Longo, M. (2006). Buying a young index: A new wrinkle on familiar strategy. *Traders Magazine*.

Ma, C. K. & Rao, R. P. (1988). Information asymmetry and options trading. *The Financial Review, 23*(1).

Maddux, J. E., & Rogers, R. W. (1983). *Protection motivation and self-efficacy: A revised theory of fear appeals and attitude change. Journal of Experimental Social Psychology. 19 (5)*.

Mao, J. C. T. (1970). Survey of capital budgeting: theory and practice. *The Journal of Finance 25*.

Matsumoto, D. A. (2002). Management's incentives to avoid negative earnings surprises. *The Accounting Review, 77*(3).

McMillan, L. G. (2002). *Options as a strategic investment* (4th ed.). New York NY: New York Institute of Finance.

McKeon, R. (2013). Returns from trading call options. *Journal of Investing, 22*(2).

Mello, A. S., & Neuhaus, H. J. (1998). A portfolio approach to risk reduction in discretely rebalanced option hedges. *Management Science, 44*(7).

Merton, R. C. (1973). Theory of rational option pricing. *Bell Journal of Economics and Management Science, 4,* 141–183.

———. (1980, December). On estimating the expected return on the market: An exploratory investigation. *Journal of Financial Economics, 8*(4).

Mukherjee, S. (2016). *The gene.* New York NY: Scribner.

Nam, C. (2011, August). Essays on a rational expectations model of dividend policy and stock returns. Dissertation paper. *Office of Graduate Studies of Texas A&M University.*

Natenberg, S. (1994). *Option volatility and pricing.* New York: McGraw-Hill.

Nayak, K. M. (2008, October 22). A study of random walk hypothesis of selected scrips. *ASBM Journal of Management, 1*(1).

Nietzsche, F. (1878). *Human, all too human.* Man and Society, #315.

Nissim, B. D., & Tchahi, T. (2011). An empirical test of 'put call parity.' *Applied Financial Economics, 21.*

Parsons, K., McCormac, A., Butavicius, M., & Ferguson, L. (2010). Human factors and information security: Individual, culture and security environment. Australia Government, Department of Defence. Command Control, Communications and Intelligence Division, Defense Science and Technology Organisation, Edinburgh, Australia.

Portema, Pablo Triana (April 8, 2008). Wither Black-Scholes? *Fortune.*

Reehl, C. B. (2003, March). Covering up with options. *Futures, 32.*

———. (2005). *The Mathematics of options trading.* New York: McGraw-Hill.

Rogers, J., Van Buskirk, A., & Skinner, D. (2009, August 17). Earnings guidance and market uncertainty. *Journal of Accounting and Economics* 48.

Rogers, R. W. (1975). A protection motivation theory of fear appeals and attitude change. *The Journal of Psychology, 91.*

Roll, R. (1977). An analytic valuation formula for unprotected American call options on stocks with known dividends, *Journal of Financial Economics, 5,* 251–258.

Roll, R., Schwartz, E., & Subrahmanyam, A. (2010, April). O/S: The relative trading activity in options and stock. *Journal of Financial Economics, 96.*

Shanthikumar, D. M. (2012). Consecutive earnings surprises: Small and large trader reactions. *The Accounting Review, 87*(5).

Sharpe, W. F. (1964, September). Capital asset prices: A theory of market equilibrium under conditions of risk. *Journal of Finance, 19*(3), 425–442.

Shiller, Robert J. (1989). *Market Volatility.* Cambridge MA: The MIT Press.

Siegel, J. (2007). *Stocks for the long run* (4th ed.). New York: McGraw–Hill.

Sitkin, S. B., & Weingart, L. R. (1995). Determinants of risky decision-making behavior: a test of the mediating role of risk perceptions and propensity. *Academy of Management Journal, 38*(6).

Sprenkle, C. (1961). Warrant prices as indicators of expectations and preferences. *Yale Economics Essays, 1*(2), 178–231.

Stäheli, U. (2013). *Spectacular speculation: Thrills, the economy, and popular discourse*. Stanford, CA Stanford University Press.

Starr, C. (1969). Social benefit versus technological risk. *Science, 165* (3899).

Stigler, S. M. (1997). *Regression toward the mean, historically considered. Statistical Methods in Medical Research, 6*(2).

Stoll, H. R. (1969). The relationship between put and call option prices. *The Journal of Finance, 24*(5).

Stoll, H. R., & Whaley, R. E. (1987). Program trading and expiration-day effects. *Financial Analysts Journal, 43*(2).

Story, L. (2011, August 17). U.S. inquiry is said to focus on S&P ratings. *New York Times*.

Strong, R. A., & Dickinson, A. (1994, January/February). Forecasting better hedge ratios. *Financial Analysts Journal, 50*(1).

Taversky, A. & Daniel, K. (1981, January 30). The framing of decisions and the psychology of choice. *Science, 211*(4481).

Treynor, J. L., & Mazuy, K. K. (O (1966, July–August). Can mutual funds outguess the market? *Harvard Business Review*, 131–136.

Tirole, J. (1982, September). On the possibility of speculation under rational expectations. *Econometrica, 50*(5).

Tobin, J., & Golub, S. A. (1997). *Money, credit and capital*. New York: McGraw-Hill.

Tokic, D. (2014). Legitimate speculation versus excessive speculation. *Journal of Asset Management, 15*(6).

Toolson, R. B. (1994). Investing in after-tax-deferred assets: A guide to determining after-tax returns. *Journal of the American Society of CLU & ChFC, 48*(5).

Vejendla, A., & Enke, D. (2013). Performance evaluation of neural networks and Garch models for forecasting volatility and option strike prices in a bull call spread strategy. *Journal of Economic Policy and Research, 8*(2).

Whaley, R. E. (1981). On the valuation of American call options on stocks with known dividends. *Journal of Financial Economics*, 9, 207–211.

Wilkens, S. (2007). Option returns versus asset-pricing theory: Evidence from the European option market. *Journal of Derivatives & Hedge Funds, 13*(2).

Yates, D. S.; Moore, D. S., & Starnes, D. S. (2003). *The Practice of Statistics* (4th ed.). New York: Freeman.

Zhang, D. (2003). Intangible assets and stock trading strategies. *Managerial Finance, 29*(10).

Index

A

Abnormal distribution 232, 234–237
Adjusted basis 118, 119, 159
Alamo Group (ALG) 20
Alphabet (GOOG) 211, 214, 227, 229
Amazon (AMZN) 147, 148
Anheuser-Busch (BUD) 202
Annualization
    Covered calls 119
    Net profit 109
    Rate of return 112
    Return if unchanged 114
    Simplified 122
    Stock returns 122
    Total return 90, 116, 118
Apple (AAPL) 138, 157
Applying basics to manage trades 280

B

Bachelier 270
Basic options quantification 279
Bear call credit spread 168, 170
Bear put debit spread 167, 168
Bell curve 12, 17, 233, 234
Bid/ask spread 38, 60, 256
Black, Fischer 258, 261, 264, 267
Black-Scholes pricing model (B-S)
    Belief in 55
    Concept 257–259
    Dividend yield and 93
    Exclusion of dividends 92
    Flaws 7, 255, 260, 264, 270
    Formula 258, 270, 278
    Implied volatility (IV) and 66
    Inaccuracies 267, 269–271
    Myths 270
    Risk-free interest rate 37
    Segments in implied volatility 37
    Use of mid-price values 38
    Variables 273
Boeing (BA) 132, 134
Bollinger Bands (BB)
    Bandwidth 75
    Buffer zone 234
    Expansion 227, 271
    Middle band 18
    M top 20
    Outer bands 228, 237
    Probability matrix 17, 18, 44, 263, 279

326 Index

Resistance tracking 18, 19
Setting goals and objectives with 24
Squeeze 22, 23
Standard deviation 18, 20, 67
Support tracking 19, 279
Visual expression of probability 237
W bottom 20, 21
Boness, James 270
Breakeven calculation
   Bear call credit spread 170
   Bear put debit spread 168
   Bull calendar spread 188, 189
   Bull call debit spread 164, 166
   Bull put credit spread 163
   Condor 172
   Covered call 149
   Covered put 156
   Covered straddle 206
   Iron butterfly 180
   Iron condor 173, 174
   Long butterfly 175
   Long call 131
   Long gut strangle 215
   Long put 134
   Long straddle 200
   Long strangle 210
   Long strap 224
   Long strip 220, 221
   Ratio write 150
   Short butterfly 179
   Short gut strangle 219
   Short straddle 203
   Short strangle 212
   Synthetic long stock 183, 184
   Synthetic short stock 185
   Uncovered call 137, 139
   Uncovered put 143
   Variable ratio write 154
Breakeven return 100–106
Bristol-Myers Squibb (BMY) 170
Buffett, Warren 264
Bull calendar spread 188–190
Bull call debit spread 164, 166
Bull put credit spread 163

Bureau of Statistics 101

C

Calendar spread 187–194
Capital Asset Pricing Model (CAPM) 271–275
Capital gains 95, 96
Cash flow to debt ratio 82
Chicago Board Options Exchange (CBOE) 110, 111
Chipotle (CMG) 191
Coca-Cola (KO) 83, 84
Collateral requirements 110, 111
Conditional probability 250, 251
Condor 172–175
Confirmation bias 8
ConocoPhillips (COP) 87, 88, 90, 121, 122
Consumer Price Index (CPUI) 101
Contingent purchase 191
Continuous random variables 15, 19
Covered call
   Comparison to uncovered puts 238
   Dividend timing 145, 146
   Earnings surprise effect 147, 148
   Losses 117–119
   Payoff calculation 143, 145–150
   Rate of return 111–115
   Strike selection 149, 150
Covered put 155–157
Covered straddle 206, 207
Cromwell's Rule 250, 251
Cultural theory of risk 243
Cumulative annual growth rate (CAGR) 122

D

Darwin, Charles 245
Debt capitalization ratio 41, 82, 83
Debt/equity ratio 82
Debt ratio 82
Decision risk 246, 247
Delta 68–70, 280

Desirable risk 241
Diagonal spread 194, 195
Discrete method 91
Disney (DIS) 236
Distribution 12
Dividend
   Achievers 41, 81
   Announcements 34
   Black-Scholes and 80, 91
   Calculations 93
   Correlation to options valuation 82
   Earnings surprise effect 146–148
   Fundamental nature of 80–87
   Lumpy 91, 92
   Payout ratio 93
   Timing 146
   Total return 79, 81, 84, 86–89, 91, 280
   Trends 35
   Yield 93, 117
Dodd-Frank Act 28
Dot.com bubble 26
Dow Chemical (DOW) 104

E

Earnings surprises 34, 146, 147, 231
Eastman Kodak (EK) 241
Economic indicators 32
Effective tax rate 101
Efficient market hypothesis (EMH) 24
Einstein, Albert x
Excel formula 37
Expanding basics to define profit and loss ranges 280, 281
Expected return 80, 81, 105, 106, 269–271
Exponential moving, average 276, 277
Exxon-Mobil (XOM) 87, 88, 90

F

Facilitation 257
Fair Disclosure Regulation 146, 147

Familiarity risk 241, 243
Fat tails 17, 234
Fear appeals 242, 243
Fibonacci retracement 242
Fluency variable 252
Fortune 269
Forward P/E 7
Framing 244
Fundamental analysis
   Correlation with technical signals 32
   Impact 31
   Intelligence gained from 32
   Options risk and 44
   Rejection of 32
   Signals 34
   Stock price and 39, 41, 44
   Trends 31
   Volatility 39, 40

G

Galton, Francis 245
Gambler's fallacy 246
Gamma 71, 72
Garman and Kohlhagen formula 262
General deterrence theory (GDT) 243
General Mills (GIS) 217, 219
General Motors (GM) 241
Glass-Steagall Act of 1933 28

H

Haug, Espen Gaarder 269
Hedging 2, 24, 70, 129
Heuristics 244
Historical volatility
   Calculation 37–39
   Consolidating range 235
   Correlation to fundamentals volatility 279
   Described 6
   Effect of news 32
   Excel formula 35, 36
   Quantifying probability 234

Reliability 7
Source 32
Trade timing 79
Hold-to-expiration assumption 125
Housing bubble 26, 28
Human nature and risk 240–242
Hypothesis for options trading 281, 282

I

IBM (IBM) 21, 22, 104, 105
Implied volatility (IV)
   Collapse 7, 8
   Compared with fundamental analysis 32
   Effect of news 32
   Extrinsic value and 63, 64
   Flaws 6, 7
   Historical volatility and 41, 45
   Manipulation 221
   Market sentiment measures 39
   Mischaracterized as predictive 66
   Option pricing and 45
   Problems with 37–39
   Realized 66
   Trade timing 79
   Value 60
Inefficiency (risk tolerance) 244
Infinite populations 252
Inflation 100–104, 106, 107
Informational efficiency 33, 66, 67
Installment calendar spread 190–194
Interest coverage ratio 82
Internet 26
Intuitive Surgical (ISRG) 75, 77
Investment, return on 110, 111
Iron butterfly 180, 181
Iron condor 173, 175

J

J.C. Penney (JCP) 33, 40, 41, 48, 50, 51
J.M. Smucker (SJM) 22
Johnson & Johnson (JNJ) 104

J.P. Morgan Chase (JPM) 121

K

Kellogg (K) 184
Keynes-Hicks theory 3

L

Leverage 131
Likelihood function 250
Liquidity 27
Long butterfly 175, 177
Long call 129, 131–134
Long gut strangle 215
Long put 129, 134, 135, 185, 186
Long straddle 200, 202
Long strangle 210, 211
Long strap 224–226
Long strip 220, 221, 223, 224
Long-term debt 32
Loss calculation
   Bear call credit spread 168
   Bear put debit spread 166
   Bull calendar spread 188
   Bull call debit spread 164
   Bull put credit spread 163
   Condor 172
   Covered call 149
   Covered put 157
   Covered straddle 207
   Iron butterfly 180
   Iron condor 173
   Long butterfly 175, 177
   Long call 131
   Long gut strangle 215
   Long put 134
   Long straddle 200
   Long strangle 210
   Long strap 224
   Long strip 220
   Ratio write 150
   Short butterfly 178
   Short gut strangle 217
   Short straddle 203

Short strangle 212
Synthetic long stock 183
Synthetic short stock 185
Uncovered call 136, 140
Uncovered put 143, 144
Variable ratio write 152
Loss recapture 25
Lumpy dividends 92, 93

M

Macy's (M) 173
Margin calculator 110
Margin Manual 110
Market maker 256, 257
Matson (MATX) 21
Merton, Robert C. 260, 262, 270
MGM Resorts (MGM) 190
Microeconomic variables 32
Moneyness 64, 73, 126, 148, 278
Monte Carlo fallacy 246
Moving average 276
M top 20
Multiple, P/E 263

N

NASDAQ 256
Net basis 149, 150
Netflix (NFLX) 150, 151, 154
New York Stock Exchange (NYSE) 45, 256
Normal distribution 12, 15

O

Objectivists (probability) 250
Occidental Petroleum (OXY) 177
Opportunity zone 237
    Analysis 32, 35
    Aspects of 237–240
    Density function 235
    Focus of 231
    Mapping system 9
    Matrix 17
    Payoff 126–128
    Risk associated with 71
    Skew factor 231, 232
    Uncertainty and 68

P

Pascal, Blaise 248, 250
Pearson's Skewness Coefficient 233, 234
P/E ratio 32, 263
Philip Morris (PM) 83, 84
Position Delta 70
Price discovery 28
Priceline (PCLN) 74, 76, 77
Pricing models 256, 281
Probability
    Additive method 4, 247–249
    Basic 3, 244, 247, 248
    Bollinger Bands (BB) 21
    Conditional 250, 251
    Cromwell's Rule 250, 251
    Danger zone 237, 238
    Discrete 248, 249
    Distribution 231
    Equally likely outcomes 237
    Interpreting 250, 251
    Likelihood function 250
    Matrix 237, 279
    Multiplicative method 5, 247, 248
    Objectivists 250
    Sample space 236, 250
    Subjectivists 250
    Theory 247–249
    Visualized 1, 9
Processing fluency 252
Profit calculation
    Bear call credit spread 168
    Bear put debit spread 166
    Bull calendar spread 188
    Bull call debit spread 164
    Bull put credit spread 163
    Condor 172

Covered call 150
Covered put 155, 156
Covered straddle 206
Iron butterfly 180
Iron condor 173
Long butterfly 175
Long call 131, 132
Long gut strangle 215
Long put 134
Long straddle 200
Long strangle 210
Long strap 224
Long strip 220
Ratio write 150
Short butterfly 178
Short gut strangle 217
Short straddle 203
Short strangle 212
Short strap 226
Synthetic long stock 182
Synthetic short stock 185
Uncovered call 137, 140
Uncovered put 142
Variable ratio write 152
Pro forma financial statements 7
Protection motivation theory (PMT) 242, 243
Proximity 9, 10
Purchasing power 103
Put/call parity 55–57, 60

R

Random variables 251
Rate of return calculations 111, 112
Rational choice theory 244
Ratio spreads 189
Ratio write 150–152
Regression toward the mean 245
Relative Strength Index (RSI) 10, 11
Retrieval fluency 251, 252
Return calculations 100, 101
Return if exercised 114, 115

Return if unchanged 114
Risk
 Articulating 8, 9, 11
 Cultural theory of 243, 244
 Decision 244
 Desirable 241
 Familiarity 241, 243
 Fear appeals 242, 243
 Framing 244
 Freeze level 130
 General deterrence theory 243
 Heuristics 244
 High versus low 11
 Human nature and 240
 Inefficiency 244
 Lost opportunity 144, 145, 149
 Market price 145
 Options and fundamental volatility 279
 Protection motivation theory (PMT) 242, 243
 Quantifying 127
 Rational choice theory 244
 Recognition 25
 Speculation 2
 Straddle 197, 198
 Theories 242–244
 Tolerance 100, 238
 Trading 126, 127
 Transfer hypothesis 3
 Variance 245
Risk-free interest rate 37
Rolled options 158, 159
Roll-Geske-Whaley 260

S

Sample space 236, 250
Sarbanes-Oxley Act 28, 221
Scholes, Myron 258
Sears (SHLD) 241
Securities and Exchange Commission (SEC) 28, 146, 147

Sharpe Measure CAPM 275
Short butterfly 178, 179
Short gut strangle 217, 219, 220
Short straddle 203, 205
Short strangle 212–214
Short strap 225
Short strip 223
Sigma (σ) 12
Southwest Airlines (LUV) 163, 166
Specialist 256
Speculation 2, 24
Sprenkle, C. 270
Standard deviation 12–15, 19, 22, 35, 44, 237
State income tax 102
Stoll, Hans 56
Straddle 57, 58, 100, 227, 229
Strangle 209–215, 218
Subjectivists (probability) 250
Supply and demand 27
Synthetic stock 56, 161, 182–184

T

Taleb, Nassim Nicholas 269, 270
Taxes 100, 102, 103, 106
Tesla (TSLA) 141, 144
Theta 65
Tiffany (TIF) 10, 11
Time value of money 100, 274
Time value premium 63, 106, 113
Total return 74, 79–81, 84, 85, 87, 88, 280
Treynor-Mazuy CAPM 274, 275

U

Uncovered call 119, 120, 137–140, 239
Uncovered put 120, 142, 143
Upper and lower bounds 62
U.S. Oil Fund (USO) 235
U.S. Treasury bonds 38

V

Variable ratio write 152, 154
Variables 9, 15, 19, 32, 126, 127
Vertical bull spread 162–167, 169, 170
Volatility
   Analysis 278
   Collapse 7, 8, 265
   Forecasting 275–278
   Key characteristic of option pricing 56
   Risk premium 67
   Surface 7
   Trends 127

W

Wal-Mart (WMT) 40, 41, 43, 46, 50, 51
W bottom 20
Weighted moving average 269

The manufacturer's authorised representative in the EU is Springer Nature Customer Service Centre GmbH, Europaplatz 3, 69115 Heidelberg, Germany. If you have any concerns regarding our products, please contact ProductSafety@springernature.com

Printed and bound by CPI Group (UK) Ltd, Croydon, CR0 4YY

23/03/2026

02076672-0018